JOHN A. SUMNER

DOMESTIC HEAT PUMPS

PRISM PRESS

Published in Great Britain 1976 by :

PRISM PRESS
Stable Court
Chalmington
Dorchester
Dorset DT2 OHB

ISBN 0 904727 09 2 Cloth
ISBN 0 904727 10 6 Paper

Printed in Great Britain by :
Unwin Brothers Limited The Gresham Press Old Woking Surrey

To my Wife - the Heat Pump Widow -
and to Stephen

Contents

TABLES

FIGURES

Foreword

This book is the summation of the author's very long interest and experience in the field of heat pumps. Well over two decades ago the author was strngly warning of the problems of the world's diminishing energy supplies, and suggesting that the heat pump was just one means of reducing our consumption of energy.

But few people took any notice. In particular the energy industry was decidedly unhelpful; for profits lay in persuading the consumer to waste rather than conserve. And as far as the heat pump itself was concerned, among other things it 'created energy from nothing', which was proof enough that it was a bogus invention.

It is, therefore, with great pleasure and a feeling of justice, that I witness this book coming to print. There is today still less excuse 'not to know', or, 'not to want to know', than ever before, and this book will, I believe, go a long way towards putting right the general ignorance that prevails.

S. D. Barnes
April 1975

Introduction

This book has been written primarily because of the great number of letters and requests for information about my heat pump installation, following references to it made from time to time by the Press in several countries.

I found that I have kept records of experiments and progress going back over twenty years and that they constitute a built-up design study for a domestic dwelling with an area of about 1500 sq.ft. The notes also constitute a diary which shows the progress in compressor and general heat pump design over nearly three decades. The question arose as to how the assembled data could be made suitable for reading and understanding by the many interested people who have stated that they have little or no engineering knowledge. I hope that this book will help these individuals, and others, to obtain at least an understanding of a domestic heat pump installation.

The book will also, I hope, clear up some misconceptions, prejudice and mystique that seem to be associated with an electric motor driving a compressor. Most of us have such an arrangement in our house in the form of a refrigerator, which abstracts heat at low temperature for our food instead of from soil or air. It lifts the temperature of that heat in the compressor and finally exhausts that higher temperature heat through a duct at the back of the refrigerator to warm up the kitchen. And, dare it be said here, the amount of heat delivered into the kitchen is always more than the heat taken from the food. When, therefore, the reader meets in the book the expression P.E.R. (Performance Energy Ratio) or the symbol R_η and, after finding that it represents the value in excess of one by which the heat put into his house by his heat pump exceeds the amount of heat (equivalent) put in to work the machine, he must not assume that he is the victim of a confidence trick. The normal refrigerator will always deliver about twice as much heat into the kitchen as is put in electrically to drive the motor, and, for a heat pump, one would expect to obtain three or four times more heat than is consumed in driving the machine. The book therefore submits two propositions for consideration and endeavours to establish their truth:

First Proposition: The method of producing a given quantity of heat by means of a properly contrived heat pump consumes only some fraction (varying from one-half to one-third) of the fuel that would be consumed by any other method to produce the same quantity of heat and will achieve the same proportional reduction in fuel cost.

<u>Second Proposition</u>: As the world supplies of fossil fuels diminish, the value and monetary saving by the use of the heat pump will increase proportionately.

Over a century ago Lord Kelvin foresaw and warned us about a world fuel shortage and devised a machine to help alleviate that shortage. This book brings his work into perspective and tries to show that pride and prejudice have attempted to hold back a useful fuel-saving method.

At this particular time a difficulty arises when writing a book for engineers. Even up to the last decade engineers have been taught to think and measure in Fahrenheit temperatures, feet and pounds (avoirdupois). As a result an engineer above thirty years of age visualizes his unit quantities in foot-pounds, whereas a physicist visualizes his unit quantities in C.G.S. or M.K.S. terms. But, since 1970, engineering students are being taught to use the Systeme International or SI units in which 1 Btu becomes 1.055 kilo-joules and 1 lb/sq.in. is shown as 6895 Newton per square metre ($6895N/m^2$). The more elderly engineer who sees the printed term $6895N/m^2$ will find little meaning in it. The same problem arises in describing refrigerants in the freon group. Even ten years ago a reference to Arcton 6, Freon 12, difluordichloromethane or CF_2Cl_2 would have been meaningful. Yet, today, the same refrigerant would be described as R12, dichlorodifluormethane with the chemical formula CCl_2F_2.

I have, therefore, adopted the somewhat dangerous course of trying to please both parties by representing a given quantity in both IS and SI systems in some cases. It is hoped that this overloading of symbols in the text will be pardoned by the younger readers and found to be helpful by older ones. There is one way in which older and younger readers may be helped - by using an approximation. SI units treat heat as a form of energy so that 1 Btu becomes 1.055 kJ. Unless detailed calculations are being made, it will only involve an error of 5.5 per cent to consider that 1 Btu = 1 kJ and, when dealing with heat flow rates, to consider 1 Btu/hr as 0.3 W.

It is expected that challenges will be made by some readers of this book and this is as it should be. I hope, however, that these challenges will be objective and related to the errors which no doubt exist or to the achievements which are alleged to have been made rather than to the eschatology of the subject.

Grateful thanks are given to my friends, Messrs Baker, Miller and Cumber for reading and making many helpful suggestions and to Tony Maufe for his invaluable help with the illustrations. Nor would I forget the loyal work of my assistants (Messrs Allwood, Clouting and Sturgeon) when building the Norwich Heat Pump.

1:   Energy supplies at AD 2000

I hope that I may be forgiven for commencing this book with an extract from one of my earlier publications. The only apology can be that we are a peculiar nation that likes to be told, from time to time, that we are in a period of crisis. We listen, and then, quite soon, either forget about it or adapt ourselves to the particular current crisis. This is probably sensible and healthy so far as the majority of us are concerned. But it is nonsensical and foolish if the minority who are in government or the 'corridors of power' refuse to publicly recognise the undoubted critical fuel crisis that we now face. Someone must have the courage to say that, if necessary, the old ways must go and new, more difficult ways must be followed. It is nearly twenty years since the warning which comprises the following section was made. Yet the electrical supply industry, for example, has done nothing during that time to examine more closely the fuel and load saving which heat pumps might make. The industry has even taken the negative step of disbanding the Heat Pump Committee of the Electrical Research Association, which made valuable progress during the twenty years of its existence.

The following section of this chapter reproduces the Chairman's Address made by me to the East Anglian Section of the Institution of Electrical Engineers in October 1956.

ENERGY SUPPLIES DURING THE NEXT FIFTY YEARS *

I am going to try to show that our long term fuel situation in Britain and Western Europe is more serious than many people think, quite apart from the immediate difficulties caused by Colonel Nasser's seizure of the Suez Canal, and that the use of heat pumps is not academic but necessary. In a society where fuel production is greater than requirements, the saving of a ton of fuel is, admittedly, of largely academic interest, but in a society where the fuel requirements are greater than the fuel available (and this is the position in Britain today), we should surely give serious thought to devoting labour and materials to the manufacture of heat pumps if it can be shown that they would achieve a substantial saving. Lord Kelvin called his machine a 'heat multiplier', a term which I want to explain later.

* Reproduced from the Journal of the Institution of Electrical Engineers, Vol. 3, April 1957.

The dissipation of Nature's Fuel Stores

For many millions of years the sun, by the process of photosynthesis, has produced something of the order of 10^{11} tons of vegetation each year, and the same amount of oxygen. During the whole of this period the vegetation and animal life died and decayed and became stored up in the form of coal, oil or peat, a growing accumulation of fuel that remained largely untapped.

Then, only two centuries or so ago, the industrial age began - first in Britain and then throughout the Western world. British forests were first denuded, and the exploitation of coal reserves followed. In Britain (and probably in the whole of Western Europe) the maximum output of coal has been passed, and a steadily decreasing annual coal output is inevitable in the future. In the Western world as a whole we used about 1 million tons of oil in 1900; today this figure is nearly 1000 million tons per year. So, in Western Europe in general and Britain in particular, we face a period of rapid growth in fuel requirements at a time when our output of indigenous fuel is decreasing. Western Europe has ceased to export energy since 1929. In a competitive world we are becoming more and more dependent upon imported oil and the long-term promise of nuclear energy. Remarkable achievements have been made by the Atomic Energy Commission and the Central Electricity Authority in developing atomic power stations, but the question is whether these can be built at a rate sufficient to close the gap between declining fossil fuel output and the increasing demand for energy throughout the world. The graph (Fig.1) shows the estimated position for Western Europe as a whole during the next twenty years - prepared by the experts of the O.E.E.C. Very

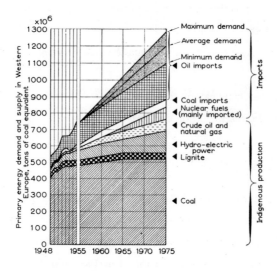

Fig. 1. Estimated fuel consumption in Western Europe 1945-75. (courtesy of the O.E.E.C.)

roughly, the same curve applies to Britain if the vertical scale is reduced by two-thirds and the contributions due to lignite and hydroelectric power are deleted. All the indications are that nuclear energy may close the gap early in the twenty-first century, but that a critical position may arise during the interim period represented by the latter half of the twentieth century.

4

The Economic Value of Saving Fuel

The problem which I am bringing to your notice therefore, is that of the possibility of a serious 'fuel gap' which could arise during the period between the decline of our native coal supplies and the possible rise of atomic power early in the next century.

The demand for oil in the USA towards the end of the century may prove to be greater than the whole world's output of oil, and there are reasons, some of which are all too apparent at the moment, for thinking that oil is likely to be a precarious source of fuel in the coming years so far as Britain, and even Western Europe, is concerned.

Capital expenditure which will save a ton of fuel annually is probably more valuable than the same amount of capital expenditure on the production or distribution of fuel, and it would appear that more attention should be given to this aspect of the question. The growing use of electricity in place of coal for domestic heating purposes, etc., provides many advantages in the form of cleaner atmosphere, the preservation of buildings and the reduction of inter-urban transport. But it does not result in an appreciable saving of coal

It has been estimated recently that Britain can afford to spend about £100 in capital equipment for each ton of fuel saved annually. Sixty-five million tons of coal are used each year in Britain in domestic dwellings. If, by some miracle, we could save 40 million of those tons each year, we could expend £4000 million in providing fuel-saving apparatus.

This obviously leads to the question: can the electricity industry take over this load and find a means of supplying the same domestic heating requirements by burning only one-half or one-quarter of the amount of coal? I think it can, and I shall suggest how this can be done. But before doing so, let me summarise the position:

a) The world demand on its stored fuel supplies is becoming incredibly greater each year, and we do not know how much of the store is left.

b) A significant contribution from nuclear energy lies in the future and its rate of growth is uncertain.

c) Britain is in a singularly precarious condition because her supply of indigenous fuel is now much less than demand.

d) A change from coal to electricity supply for domestic heating in Britain would make no contribution to her serious fuel problem, unless the industry can find a means of making one kilowatt-hour of electricity, when used for heating, do as much work as, say, three or four kilowatt-hours at present.

The Heat Pump

If we compress air, its temperature will rise, and heat at a higher temperature will become available. May I may it quite clear that the heat obtained under compression is exactly equal to the amount of work done in compressing the air. So far, therefore, we have only obtained an amount of heat exactly equivalent

5

to the amount of work done.

Now, if we wait long enough, the higher-temperature heat of the compressed air will be given up to the room. But we still have a volume of air in the cylinder which, although cooled, is at a higher pressure than when we started and is therefore capable of doing further work when we let it expand back to atmospheric pressure. In fact, it may be enough to do three-quarters of the original work of compression if we feed it into another cylinder which is coupled to the original compression cylinder. In this way, therefore, we should have obtained an amount of heat for only a quarter of the work done in providing that heat, or, to put it another way, one kilowatt-hour of electricity used to drive the machine will give the same amount of heat as four kilowatt-hours of electricity used for heating in the orthodox manner. One could put this in the form of an equation:

$$
\underset{\substack{\text{Heat at high} \\ \text{temperature}}}{4 \text{ kWh}} = \underset{\text{Work}}{1 \text{ kWh}} + \underset{\substack{\text{Heat at lower} \\ \text{temperature}}}{3 \text{ kWh}}
$$

Now we know that we cannot create energy, and so where did this extra 3 kWh of heat come from? Obviously it is heat that was already present in the air at a lower temperature than the room, and which has been up-graded to the higher temperature sufficient to warm the room. The heat pump, in fact, draws heat from a low temperature source and pumps it into a higher temperature sink.

The Effect of Heat Pumps on the Nation's Heating Budget

The heat energy available from 65 million tons of coal at 30 per cent efficiency is equivalent to 130 million MWh. This is nearly twice the present total annual output of all power stations.

This heating load, at 10 per cent annual load factor, represents plant capacity of about 130,000 MW or about eight times the present available capacity in all stations. If heat pumps were used with a Performance Energy Ration of 3.5 and an annual load factor of 30 per cent, a reduction from 130 million MWh to 37 million MWh per annum would result, and the necessary plant capacity would be 11,000 MW instead of 130,000 MW. The coal consumption at the power stations would be 18.5 million tons per annum instead of 65 million tons - a saving of 47 million tons of coal each year.

We could, therefore, as a nation, spend £100 x 47 million tons, i.e. £4700 million, in capital equipment on heat pumps in place of expenditure on machinery and facilities to bolster up a declining supply of fossil fuels.

May I suggest that I have shown a way in which the electricity supply industry could make a substantial contribution to the serious fuel problem that is likely to face Britain during the rest of this century.

Conclusion

A scheme of such magnitude would require the wholehearted support of the industry. There would be a need on the part of all the Boards and of the Central Electricity Authority to overcome prejudice and to examine the potentialities of the heat pump. No one person can do this alone, and at least one Board must put its resources behind the enquiry. To foster public interest the concentration of the Boards' immense selling organisation should be behind the scheme, with or without the co-operation of the manufacturers.

Nature has given us illimitable sources of prepared low-grade heat. Will human organisations co-operate to provide the machine to use Nature's gift?

THE FUEL POSITION IN 1976

How closely does the fuel demand and supply position correspond, in 1976, to the estimates of 1956? So far as demand is concerned, the overall estimate is fairly accurate even though the component parts are somewhat different. The estimates as to coal and nuclear fuels used are somewhat low, and to 1973 the demand for oil and natural gas was higher than estimated. Up to 1965 the total value of new oil reserves found each year was in excess of annual consumption, but since then, despite North Sea finds, this condition no longer prevails. By 1973 there was some evidence that the annual demand for oil was increasing so rapidly that it was likely that total world supplies of oil might be exhausted by the end of this century (including North Sea supplies). Unless the appalling annual rate of increase in the use of fuel is reduced, the twenty-first century will certainly see the end of the world's oil supplies. I can find no published figures on the amount of uranium in the world but it must be a finite amount.

The consequences which must follow the possibly imminent termination of any one of our fossil fuel supplies are radical in their nature. Unless a proved substitute is available - and this occurs usually after many decades of experiment and much capital investment - the life, even the survival, of nations is at risk.

2:   Historical development of the heat pump

The first mention of a 'heat multiplier' is associated with William Thomson (later Lord Kelvin) as a part of his great generalisation - the theory of the dissipation of energy. In several papers on this subject he pointed out that motive power was obtained only by 'degrading' heat, i.e. burning fuel, and that the heat thereby rejected represents energy dispersed and 'unavailable' as further motive power. Kelvin anticipated a time when all fuel would be exhausted and motive power therefore no longer available. As one of the world's first conservationists he therefore outlined and designed a machine which he called a Heat Multiplier. This machine would permit a room to be heated to a higher temperature than the ambient temperature, by using less fuel in the machine than if such fuel was burned directly in a furnace.

So far as I can establish, a machine for heating buildings using Kelvin's cycle and specification was never built. Nor, apart from the standpoint of scientific research was there any reason why one should be built. Coal and wood were available in abundance and future supplies seemed guaranteed. Also, from a review of the contemporary history, there was no demand for the comfortable heated dwelling in winter that we demand today. The only interest that did arise immediately in Kelvin's machine was for the cooling of public buildings and British residences in India.

Heat Pump Development in Britain

Despite Britain's highly critical fuel position, there has been no appreciable development in heat pumps in the 120 years that have elapsed since 1852, unlike in the USA where the number of domestic heat pumps used for either room heating or cooling exceed one million. I stayed recently in a relatively small hotel in New York which had over 400 heat pumps in regular use. The occupant of the suite was able to turn the machine to heating or cooling, as desired.

So far as I know, the first two heat pumps built in Britain were the machine built by J.G.N.Haldane in 1930 as a small domestic machine and the much larger commercial machine built by myself in 1946 which I refer to later.

In Britain three promising starts were made soon after the Second World War. With fuel shortages striking home, a small group of people decided to do something positive about the situation. A small budget and facilities were provided at the Electrical Research Association laboratory at Leatherhead, Surrey and research work commenced on the heat pump under the guidance of a small technical committee drawn from the electrical manufacturing and supply industries and involving government and other scientists.

Due to the enthusiastic work of Miss M.Griffiths, a feasibility study was made to consider possible applications of the heat pump in Britain. Investigations were devoted to a study of the simple reversed vapour compression cycle. A long series of practical experiments were made to establish likely Performance Energy Ratios, with sources of low-grade heat such as soil, air and water, and a number of valuable publications were published giving the results of the committee's work. The committee was disbanded when the work of the Electrical Research Association was taken over by the Electricity Council.

In 1948 the late Lord Nuffield engaged me as a consultant and twelve prototype heat pumps, each with a 3 kW motor drive, were made and installed for testing for two years in selected houses. Ground coils were used as the low-grade heat source. The results of the tests were very satisfactory as the heat input to the house averaged 9 kW, giving a P.E.R. of 3:1. The project ceased in 1950 because of administrative hesitation in setting up the large organisation required to form a country-wide sales and service network.

Some years later the Lucas Company decided to devote an even larger sum of money to develop domestic heat pumps, using an air-to-air system. The company developed and manufactured an admirable and efficient hermetically sealed compressor unit, and, on my advice, built two test houses with full recording apparatus. In order to meet the problem of sales and servicing, the company were advised that such an organisation already existed, country-wide, in the form of the Electricity Boards. They had both the labour and the expertise to handle the task.

After two winters of testing the two prototypes the company compiled test results which were forwarded to each Electricity Board with an invitation to attend a conference. This invitation was accepted by one Board only and ultimately this third promising scheme was abandoned.

A brief historical note is given here relating to two of the larger industrial heat pumps constructed between 1945 and 1955. In 1952 a heat pump was designed and constructed for London's Festival Hall. It was not successful largely due to errors in design. The unusually high standard of insulation of the Festival Hall due to soundproofing, and the large heat emission of the occupants, do not appear to have been taken into consideration. The result was that the heat pump output was too great to allow the machine to be run except for short periods. A special type of centrifugal compressor was used which had to be run at high speeds and thereby created a serious high-frequency noise problem. The well-publicised failure of this machine would appear to have prejudiced heat pump development in recent years.

The Norwich Heat Pump* was designed and built by me for the Norwich Corporation Electricity Department in 1945. During its life it had the effect of eliminating the consumption of 136 tons of coal per year at £3.25 delivered to the building, by substituting the consumption of 77 tons of coal per year delivered to the

---

* The Norwich Heat Pump, Journal of the Institution of Mechanical Engineers, Vol.158, No.1, June 1948.

power station. The Corporation Electricity Department (and the heat pump) was taken over by the Eastern Electricity Board on nationalisation. The Commercial Heating Section of the Board viewed with some disfavour a machine providing a unit consumption of only one-third of that required to provide a given amount of heat by normal resistance heating. Accordingly, despite a relatively long period of satisfactory running, the Board dismantled and destroyed the plant.

Many of the reasons put forward against development are the standard ones which apply to most new ideas, particularly those which are not easily understood, e.g. :

a) The capital cost is too high.

b) There is no proven fuel saving.

c) Refusal to accept that a machine can deliver more heat than the equivalent work input.

d) Since rotating machinery is involved, breakdowns may be frequent and maintenance costs high.

With regard to the first two points: my original installation, including garden coil, was valued at £300 in 1952. A further £100 capital was spent in 1962. Allowing a 20-year life (already expired), this represents an annual cost of £20. The annual cost of electricity for driving the plant and heating the premises averaged £50 per year to 1970 as compared with £130 per year if resistance elements had been used to heat the house and hot water. The capital cost of the installation was amortised in 1972, resulting in an accumulated sum subsequently saved of £1300 towards future replacements.

The third point needs little comment except to say that jocular scepticism is a pleasant feature of life. A statement made by a university engineering lecturer, that the description given by Lord Kelvin is impossible to accept, or by another lecturer, that a heat pump offends thermodynamic law, is a less pleasant feature.

The fourth point is simple to answer in that all maintenance work since 1952 has been carried out by a refrigeration engineer employed by the Eastern Electricity Board, trained by myself in heat pump operation. This man's time and all materials used have been charged by the E.E.B. at normal commercial rates. The total annual cost of maintenance has averaged £6.

The Likely Effect of National Development of Domestic Heat Pumps

If the development of domestic heat pumps by the Lucas Company had not been thwarted, probably some half a million installations would now be in use. It was estimated then that if annual sales of heat pumps equalled the annual sales of refrigerators, then heat pumps could have been sold for £400, inclusive of the fuel tax on electrically driven heat pumps which the government imposed at the time.

The electricity supply industry at that time had experienced two occasions when scientific improvements in components had adversely affected, if only temporarily, the sales of electricity. The first experience (1910-1920) came when carbon filament lamps were superseded by metal filament lamps, thus reducing consumption on lighting by about two-thirds but greatly compensated for by increased sales. A less dramatic reduction in consumption occurred when gas-filled tubes replaced metal filament lamps, but again only temporarily. If the Lucas Company's development had materialised, when the national electric house heating load was probably 25 per cent of the total load, there would have been, over the next few years, a loss of revenue and of units sold of the order of $12\frac{1}{2}$ per cent. This would not have been welcomed. However there would also have been a considerable saving in cable, transformer and transmission line capacity to be offset against the reduction in units sold. Each consumer with whole-house electrical heating would have reduced his demand on this plant from 13 kW to 4 kW thereby increasing the life of the plant by nearly three times. The reduction in plant demand would doubtless have been welcomed by the C.E.G.B., who at that time were barely able to meet the national peak demand, and the subsequent enormous and expensive increase in new generating plant which they had to meet would have been delayed, or even partially avoided. In fact, it might have been cheaper to subsidise the half-million new heat pump owners with £400 each than to pay the £200 million that might have been saved on distribution, transmission and generating plant, and the cost of the 5 million tons of coal that might have been saved each year at the generating stations. Actually, the greater danger was that non-consumers using oil heating might have changed to the cheaper method of electric heating by heat pump.

I suggest that there should be reconstituted a new government committee to consider the use of electrically driven heat pumps on which manufacturers and the C.E.G.B. should have a major voice, with the Electricity Council available to provide sales and service facilities and to carry out directed experimental work.

There is, of course, a considerable market for domestic heat pumps driven otherwise than by electricity. The most promising development is likely to be the use of an oil engine working on the Stirling cycle driving a similar engine on the reverse cycle.

3:   First principles

General Principles

As an introduction to this chapter a brief description is given of the terms 'heat', 'temperature' and also 'energy'; some knowledge of the manner in which these terms are used later in the book is essential to any analysis of heat pump operation.

The heat which a body contains is measured by the energy of motion of all its particles. To increase the temperature of a body is to increase this energy and lowering the temperature will decrease it. There- fore, temperature is a measure of heat energy. If two bodies at different temperatures are in contact, heat from the body at the higher temperature will be transferred to the body at the lower temperature until the two bodies are at an equal temperature. The temperature (heat energy) of a body cannot be increased ex- cept by adding energy to it; the heat pump does this in the form of work by compression.

Energy can be expressed as a capability for doing work and, to do work, it is essential that there be a temperature difference. The quantity of energy possessed by a body is measured by the amount of work it can do, which again, is measured by the temperature difference that it can create. But, despite the changes of energy which may occur in a given system of bodies, the total energy of the system remains unaltered.

A rock weighing one ton, poised on the edge of a cliff, possesses potential energy. If the rock falls to the ground below, the energy lost in falling will be given, partly to the air as it falls, but mainly to the object which it strikes when it comes to rest. The rock may then be restored to its initial energy state by adding energy given by machines or men to lift it back to its original position. A pound of coal, oil or wood possesses a given amount of such potential energy. If combustion takes place in the presence of oxygen, chemical and heat energy is released and the original energy of the fuel will fall by an equal amount. But in this case neither machines nor men can ever restore to the fuel the value of the 'high-grade' energy which it contained originally.

When fuel is burned, heat energy is released at a high temperature. If the fuel is burned directly, as in a grate, the heat energy will be dissipated into air so that the temperature continues to fall until it eq- ualises with the ambient temperature and the temperature can fall no more. The ambient surroundings will have absorbed the 'high-grade' energy initially 'Available' in the fuel. Alternatively, the heat energy in the fuel may be released, say, in a boiler in which the water absorbs some of the energy so that it becomes heated and turns into steam at varying degrees of pressure. The heat energy of the steam may turn an engine or turbine so that some part of the heat energy can, in turn, be converted into mechanical (work) energy. But the energy available from the work and that remaining in the exhausted steam must inevitably fall in temperature until it too reaches that 'Unavailable' ambient temperature to which all released 'Available' energy must fall.

It is fortunate that we all live in this 'bath' of ambient heat, which, in Britain, varies in temperature from about 0^oC to about 42^oC. All the heat generated by combustion ultimately sinks into this 'bath' and if it did not exist the heat would continue to fall in temperature until it reached absolute zero (-273^oC) whereupon the heat would be both devoid of temperature and energy.

It is the presence of this surrounding cushion of 'Useful' heat that makes the operation of the heat pump possible. The reader will, at first, question whether heat at or near freezing point temperature can properly be classified as 'heat'. This is because experience has led us to use an artificial scale to determine heat energy in which the sense of touch uses blood or body temperatures of 34.4^oC as the touchstone. Water freezes at 0^oC and turns into steam at 100^oC and these limiting changes of state tend to indicate our limit of 'coldness' as being 0^oC, or freezing point. In fact a body is only 'cold', i.e. has no energy or temperature, when it is 273^oC below freezing point. Scientifically, and practically, when considering heat pumps it is necessary to use a different, or 'absolute' scale of temperature. This is shown in ^{o}K instead of ^{o}C and has equivalent values as follows:

$$0^oK = -273^oC = \text{absolute zero}$$
$$273^oK = 0^oC = \text{freezing point of water}$$
$$373^oK = 100^oC = \text{boiling point of water}$$
$$\text{i.e. } ^oK = {}^oC + 273^o$$

Therefore, a given quantity of a substance at freezing point temperature must be considered as having 273 units of heat energy, i.e. as being 273^oK 'hot' and at boiling point as having 373 units of heat energy, and so on.

For the reasons given above, it will be clear that the rapidly decreasing supply of 'high-grade', 'Available' energy is being dissipated, not destroyed, or reduced in quantity, and is being absorbed and locked up in a rapidly increasing store of 'low-grade','Unavailable' heat in our environment. The question is – how can we use this huge potential asset? The purpose of this book is to show that we can, and must, dip into this store of ambient energy before it is too late. Let us review the foregoing principles by asking a question.

Why should I use a heat pump?

The short answer is that you require heat which, in winter will be at a temperature higher than that of the ambient air surrounding the room and sometimes, in summer, heat which will be required at a lower rel-

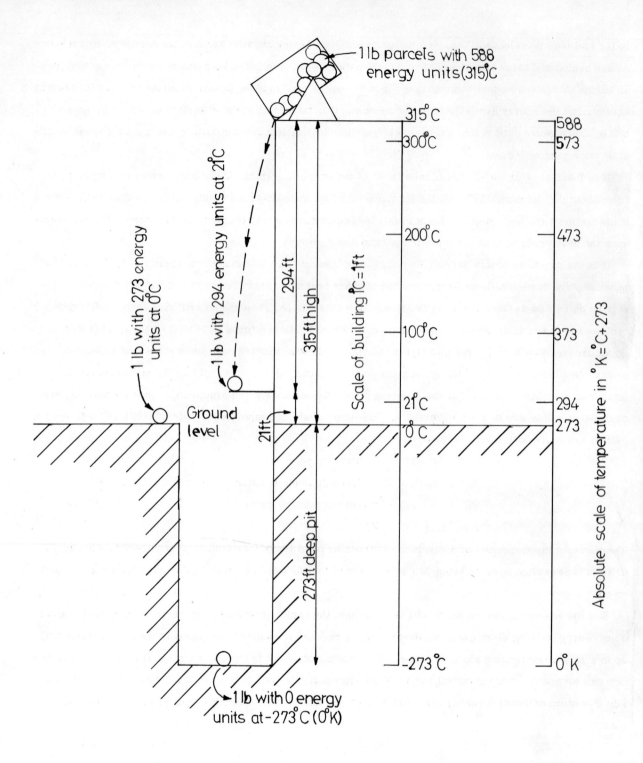

Fig. 2. Diagrammatic representation of the two methods of obtaining heat.

ative temperature. The first requirement is most easily met by striking a match to cause coal or oil to be burned by a chemical process which produces heat at, say, 315°C. The heat thus made available must then be allowed to fall in temperature to about 21°C to heat the room. As it continues to fall the heat will escape from the room and be absorbed by the environment at about 0°C as we have seen.

Nature imposes penalties in the processes of the creation and usage of heat. Firstly, the 1 lb of coal or oil from which the heat was derived, and which took millions of years to make, has disappeared in a matter of minutes. Secondly, she decrees that we can never obtain, even in the most efficient circumstances, more than about 80 per cent of the potential heat in the fuel. For most domestic combustion processes we obtain about 50 per cent of this potential heat. The remainder escapes via the chimney and as unconsumed fuel into our ambient surroundings. So, as an average, in order to obtain the heat at 21°C, available in 1 lb of fuel by direct combustion, 2 lbs of fuel must be destroyed and the total heat from these 2 lbs will ultimately be absorbed by our ambient surroundings.

Returning now to the 'bath' of ambient heat that surrounds us, it will be seen that a given quantity of, say, air or water always contains from about 273 to 315 units of heat energy, i.e. it is at a temperature of about 273°K (0°C) to 315°K (42°C). If we mix an equal quantity of air or water at the two limiting temperatures shown above, the resultant temperature would be at 294°K (21°C), which is the normal winter room heat temperature. So, instead of striking a match to obtain heat, an alternative method which now suggests itself is to start by taking a given quantity of air or water which is at, say, 273°K (0°C) and which already contains 273 units of heat. Then to find a means of adding 21 more heat units so that we finish the process with the air or water possessing 294 units of heat at a temperature of 294°K (21°C).

The difference between the direct combustion and the heat pump processes will now be evident. The direct combustion process is one of subtraction in which we subtract and waste 294 units from fuel which contains, say, 588 units (at 588°K or 315°C) to obtain 294 units. By comparison the heat pump process is one of addition in that we commence with a quantity of air or water containing 273 units and then add to it by some means 21 units only, to obtain the required 294 units.

Figure 2 shows a scientifically inaccurate but practical way of showing the way in which the two different methods are used to obtain heat at 294°K (21°C). The heat energy in a given parcel of fuel is assumed to be proportional to the height of the fuel above the bottom of the pit which is 273 ft below ground level. A parcel of fuel at the top of the building 315 ft high (or 588 ft above the bottom of the pit) will therefore contain 588 energy units. If this parcel is allowed to fall 294 ft onto a platform 21 ft above ground level, i.e. at 294°K (21°C), only 294 energy units become available. This process of falling, representing direct combustion at 50 per cent efficiency, is assumed to result in an energy loss of 294 units.

With the heat pump, we first take a parcel of fuel already containing 273 units, from a huge store lying at ground level (273°K or 0°C) and do compression work representing 21 energy units so as to lift the parcel 21 ft to platform height so that it, thereby, contains $273 + 21$ or 294 units of energy. In the direct combustion process we lose 294 energy units and in the second case we provide only 21 units to achieve the desired

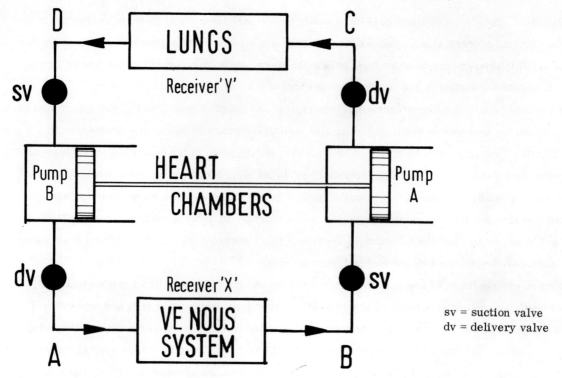

Fig. 3a. Analogue of human heart.

sv = suction valve
dv = delivery valve

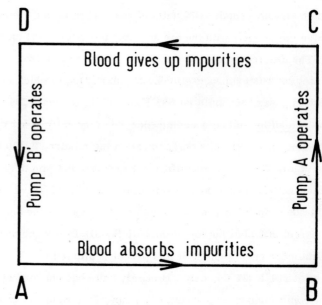

Fig. 3b. Cycle of blood flow in human body.

In Fig. 3a the blood flow system of the human body is represented diagrammatically. Each pump operates seventy times per minute during which time they transfer from 1 to 5 gal of impure blood between the receivers X and Y. Blood which has taken in impurities during operation AB in Fig. 3b is pumped from receiver X by pump A during operation BC to enter receiver Y so becoming purified during operation CD. The cycle is completed when pump stroke DA restores the newly oxygenated blood to receiver Y. Normal operation comprises 100 million cycles.

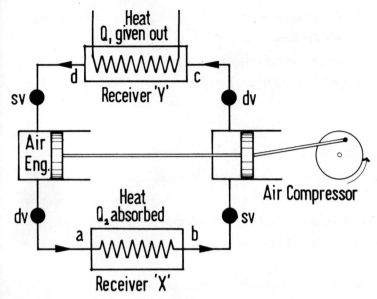

Fig. 3c. Mechanism of air-type heat pump.

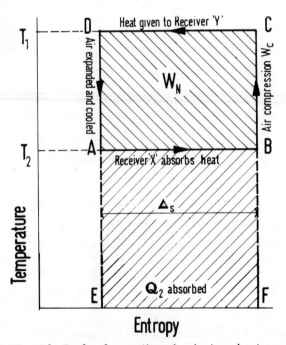

Fig. 3d. Cycle of operations in air-type heat pump (Carnot ideal cycle).

In an ideal heat pump, as represented in Fig. 3c, the components and operations are very similar to those in Figs.3a and 3b. In Fig. 3c the mechanism uses air as the 'working substance' being repeatedly transferred. Fig. 6 has a volatile liquid replacing blood as the 'working substance' or 'refrigerant' being pumped at a rate (for the normal domestic heat pump) of approximately 1 gal per minute (0.63 l/min). Fig. 3c represents an ideal heat pump operating on the ideal Carnot cycle of Fig. 3d and is described more fully in Appendix B. Fig. 7b shows the less efficient and practical Joule reversed cycle upon which it would work, with cycle abcd instead of ABCD.

294 energy units. To provide 21 units of power in a compressor, at 33 per cent efficiency, we must provide 3 x 21 or 63 energy units. In this case the heat pump process is $\frac{294}{63}$ or 4.8 times as efficient as the direct burning process. It will be noted that an increase in the height above ground to which the heat pump must lift the parcel, i.e. an increase in the temperature differences between which the heat pump will work, will reduce this advantage of 4.8.

Cycles of Operation (when air is used as the refrigerant)

A heat pump works by performing repeatedly one or other of several possible fixed cycles of operations. An approximate analogy to this regular cycle of operations may be found in the working of the human heart. The heart is divided down the middle into two major parts, forming two separate pumps side by side, each half with an upper and lower chamber. The cycle of operations may be seen in Figs. 3a and 3b and are as follows:

1st operation : B to C – the pump between B and C takes in 'used' blood from the veins (receiver X) and pumps it into the lungs (receiver Y).

2nd operation: C to D – the blood changes its state in receiver Y by releasing carbon dioxide and absorbing oxygen from the lungs.

3rd operation: D to A – the pump between D and A passes the blood into the circulatory system.

4th operation : A to B – the blood from the arteries enters receiver X (the circulatory system) which again changes its state by absorbing carbon dioxide.

An analogy is a tender plant which should not be forced and, with this limitation in mind, turn to Figs. 3c and 3d which represent a diagrammatic heat pump and its cycle of operations when using air instead of blood as the working substance, and notice the comparison with Figs. 3a and 3b. The operations for the air heat pump are as follows:

1st operation : B to C – the compressor pump P_C takes in an amount of air having a heat value q_2 from receiver X at the normal ambient air temperature t_2. Work ($= w_c$) must now be supplied so that the pump compresses, thereby heating the air to the higher temperature t_1. An amount of heat ($w_{c/j}$) is thereby added to q_2.

2nd operation: C to D – air with an amount of heat q_1 at t_1 (totalling $q_2 + w_{c/j}$) enters the receiver Y where it gives up heat isothermally to the room equal to q_1.

3rd operation: D to A – air in receiver Y still retains most of its original pressure as it enters the expansion pump P_E where it gives out work $= w_e$ and expands back to the original lower pressure.

4th operation : A to B – the air falls in both pressure and temperature before it enters receiver X, where it returns to its original condition prior to the first operation. During AB the air takes in heat (q_2) at t_2 from the atmosphere.

Referring now to Fig. 3d: if we could take in the quantity of heat q_2 along line AB at temperature t_2

when the outside atmosphere is at t_2, the rectangle ABFE would represent q_2. The vertical line BC represents the <u>external work</u> $w_{c/j}$ put in to lift the air temperature from t_2 to t_1. We now have available $q_2 + w_{c/j}$ at temperature t_1, represented by the larger rectangle DCFE which in turn represents the heat q_1 at the higher temperature t_1 now given out to heat the room. But, after giving out q_1 at t_1 along CD we still have some air at the higher temperature t_1 and pressure p_1. We let this air do <u>work for us</u> = jw_e along DA so that the net work which we had to supply becomes $w_c - w_e = w_n$.

Overall, therefore, we were given an amount of heat q_2 at t_2 from the surrounding atmosphere; this is represented by the area ABFE. By adding the work of compression w_c we raised the temperature of the whole mass to q_1 at t_1 and can express the results of the cycle, up to point D as:

$$q_1 \text{ at } t_1 = q_2 + w_{c/j}$$

i.e. so far we have only obtained as much extra heat (above q_2) as we put in as work ($w_{c/j}$).

But then the expansion engine did some work for us, when:

$$q_1 \text{ at } t_1 = q_2 + (w_c - w_e) = q_2 + w_{n/j}$$

at point D, when the cycle is complete.

So that, for the complete cycle, we can consider the areas within DCFE in Fig. 3d as representing heat and work quantities and also temperatures, when:

Heat given out at high temperature = q_1 at t_1 = area DCFE.

Net work supplied = $w_c - w_e = w_n$ = area DCBA.

Heat given us by ambient air = q_2 at t_2 = area ABFE.

We can then deduce that ;

Area DCFE = area ABCD + area ABFE

or $\qquad q_1 = q_2 + w_{n/j}$.

Also $\qquad q_1 - q_2 = w_{n/j}$

or $\qquad q_1 - w_{n/j} = q_2$.

Note that an amount of work = w can be expressed as an equivalent amount of heat = q by dividing by an equivalent j = 778. Hence 1 horsepower or 33,000 ft lbs is the thermal equivalent of $w/j = \dfrac{33,000}{778} = q = $ 2545 Btu or 748 Watts. (See Appendix A)

The Efficiency of a Heat Pump

We continuously try to measure the 'efficiency' of our various operations, e.g. in currency exchange or miles per gallon. We place one gallon in the tank and find that the car runs for 20 miles giving an 'efficiency' of miles over gallons of $\dfrac{20}{1}$. The 'efficiency' of a heat pump must obviously be measured in a similar manner, i.e. the ratio of 'Heat given out' to 'Heat put in as work'.

In fact, we have already taken these measurements, first in Fig. 2 and then in Fig. 3d. In Fig. 2 it was found that 294 energy units were made available after 21 energy units of pumping work were put in,

$$\text{Efficiency (per Fig. 2)} = \frac{\text{Heat given out}}{\text{Heat put in as work}} = \frac{294}{294 - 273} = 14.$$

Then, from Fig. 3d:

$$\text{Efficiency (per Fig. 3d)} = \frac{\text{Heat given out}}{\text{Heat put in as work}} = \frac{\text{Area CDEF}}{\text{Area CDEF} - \text{Area BAEF}}$$

These two areas have an equal width of base and their ratio can therefore be expressed in terms of their height along the left hand absolute scale of temperature in Fig. 3d:

$$\text{Efficiency (per Fig. 3d)} = \frac{\text{Height of Area CDEF}}{\text{Height of Area BAEF}} = \frac{t_1}{t_1 - t_2} = R_\eta .$$

Therefore, in both cases it can be shown that:

$$\text{Efficiency of heat pump} = \frac{294}{294 - 273} = \frac{t_1}{t_1 - t_2} \quad \text{in Fig. 2}$$

and

$$= \frac{t_1}{t_1 - t_2} = R_\eta \quad \text{in Fig. 3d.}$$

There is one other important measure of efficiency that can be made from Fig. 3d. We can measure the efficiency of the machine when acting as a <u>refrigerator</u>. Then:

$$\text{Efficiency of refrigerator} = \frac{\text{Area BAEF}}{\text{Area CDEF} - \text{Area BAEF}} = \frac{t_2}{t_1 - t_2} .$$

For many years this latter efficiency has been known as the 'Co-efficient of Performance' or C.O.P., i.e:

$$\text{C.O.P. of a refrigerator} = \frac{t_2}{t_1 - t_2} \quad \text{which also} = \frac{(t_1)}{(t_1 - t_2)} - 1.$$

Reference should be made here to the name which should be applied to the term $\dfrac{t_1}{t_1 - t_2}$ about which there is still a great difference of opinion. The terms 'Advantage', 'Performance Efficiency Ratio' and 'Co-efficient of Amplification' have been used, and also 'Co-efficient of Performance'. This last term is obviously unsuitable and is numerically wrong. For a century this term has been used as a measure of the 'refrigerating capacity' of a machine, i.e.

$$\text{Co-efficient of Performance} = \frac{t_2}{t_1 - t_2} .$$

Now:

$$\left(\frac{t_2}{t_1 - t_2} \right) = \left[\left(\frac{t_1}{t_1 - t_2} \right) - 1 \right]$$

and these terms have different values; therefore they should have different designations.

For many years the 'efficiency' of an ideal heat engine doing work has been agreed as:

$$\text{Efficiency } \eta = \frac{t_1 - t_2}{t_1} \quad \text{(for a Carnot heat engine).}$$

20

Therefore the reciprocal $R_\eta = \dfrac{t_1}{t_1 - t_2}$ (for a Carnot heat pump).

In 1948 I suggested* that the term 'Reciprocal Thermal Efficiency' (denoted by R_η) should be used, i.e.:

$$\frac{t_1}{t_1 - t_2} = \frac{1}{\eta} = \text{Reciprocal Thermal Efficiency} = R_\eta .$$

In 1948 Professor S.J. Davies** referred to the important fact that we are measuring a 'ratio' of quantities of energy and suggested the term 'Performance Energy Ratio'. In my view this may be a more acceptable term. In this book, therefore, the term 'Performance Energy Ratio' has been used, allied to the subscripts of either P.E.R. or R_η .

*The Norwich Heat Pump, Proceedings of the Institution of Mechanical Engineers, Vol. 158, No. 1, 1948, p.22.
** Heat Pumps and Thermal Compressors, Constable, 1948.

The Recent Demand for a Fuel-Saving Machine

Some reference should now be made to the sudden increased interest in domestic heat pumps. But first let us reconsider the whole system of operation by making a brief reference to two fundamental laws of thermo-dynamics.

The first law states that we can always convert the whole of any given quantity of mechanical energy into its equivalent in heat energy.

The second law states that we cannot convert more than some fraction of a given quantity of heat into mechanical energy.

Even for our most efficient engines we can only obtain about 33 per cent of the heat as useful work leaving the balance of 67 per cent to be wasted into earth, air or water. The ratio

$$\frac{\text{Heat of work available}}{\text{Heat put in}} = \frac{1 - 0.67}{1} = \frac{0.33}{1} = \text{or } \frac{3}{10\text{ths}} = \text{Efficiency } (\eta) .$$

When a heat pump is used, we reverse the above process and do work so as to obtain heat, and because of this reversal we should expect to reverse the above expression, when:

$$R_\eta = \frac{\text{Heat available}}{\text{Work put in}} = \frac{1}{0.33} \text{ or } \frac{3}{1} = \text{P.E.R. } (\frac{1}{\eta}) .$$

With a heat pump, as has been shown, we can abstract two units of low-grade heat that the engine rejected, then provide one more unit in the form of work so as to lift the temperature and thereby provide three units of higher-temperature heat for each equivalent work/heat unit expended. There has never been any effective denial that the above exposition is correct, ever since it was first announced in 1852. But it was held to be cheaper and more convenient to use fuel more wastefully by direct burning. For, to use a heat pump, it is necessary to devise some means and apparatus to collect a given quantity (q_2) of low-grade heat at a temperature t_2, say, $32^\circ F$ ($0^\circ C$). Then machinery must be provided that will supply a further amount of heat in the form of the work (w_n) required to drive the machine. Then the sum of these two quantities represented by q_2 at t_2 and w_n needs to be collected and processed in a further component so that we finally have an amount of useful heat that can be represented by the following expressions:

$$q_1 \text{ (at } t_1 = 140^\circ F) = q_2 \text{ (at } t_2 = 32^\circ F) + \text{work (w)}$$

in a ratio: 3 heat units at $140^\circ F$ = 2 heat units at $32^\circ F$ + 1 heat unit of work, when $\dfrac{q_1}{w} = \dfrac{3}{1}$.

Until quite recently it has been held that the sources of high-grade heat to provide q_1 are limitless and cheap. It has been considered witless to go to the expense and trouble to provide expensive machinery and plant in order to upgrade q_2 so as to obtain q_1, when a lighted match or switch can give q_1 without further trouble. The facts are entirely different. It can be shown that the many millions of pounds spent on machinery to obtain coal and oil is a cost very much greater than that of providing machinery and components for heat pumps, which supply a given amount of useful heat, for purposes within given temperature limits with less fuel consumption than by any other means.

The evidence for this statement is provided by the relative costs of supplying 1230 therms (36,050 kWh) of heat annually for centrally heating my own house. If this quantity of heat had been supplied by electric resistance heating (storage heaters, radiators etc.), the Electricity Board would have charged £330 in March 1975. Alternatively, if 1230 therms (129.76 GJ) were supplied from good coal burned at 50 per cent efficiency, it would require 9 tons for which the Coal Board would charge about £240. Oil or gas would cost about the same as coal.

The alternative method of obtaining 1230 therms used by myself for the last twenty years (to 1975) was to spend £450 of capital on machinery and equipment that will extract from the earth 820 therms of energy at an annual cost of £25. Then to pay the Electricity Board £110 for a further 410 therms (12,000 kWh) of electrical energy which, when added to the 820 therms taken from Nature's free pool, will provide 1230 therms for £135. If the heat pump machinery cost four times as much as shown above, the annual cost would be comparable with the cost of burning 'raw' electricity, coal or oil and, above all, only 5 tons of coal need have been burned annually at the power station instead of 9 tons of coal or about 6 tons of oil. The question must arise as to whether some part of capital investment could economically be directed into providing fuel-saving appliances.

As regards the conservation of fuel, there has never been any valid doubt since 1852 that the heat pump can produce heat at $110^\circ F$ to $140^\circ F$ ($43.3^\circ C$ to $60^\circ C$) by using less fuel than any other device, with the exception of solar heat. Monetary conservation has hitherto been the primary consideration and, although the example quoted in this book proves that the heat pump has provided both fuel and monetary conservation, the example has not been sufficient to overcome prejudice, and often, lack of knowledge.

Since 1973 fuel prices have risen so much that there can no longer be any doubt that the heat pump conserves both fuel and money. Fig.4 demonstrates this fact and is based on fuel prices in March 1975. Further increases in fuel prices are imminent; but as can be seen in the Figure, they will not seriously alter the price gap shown between the heat pump and alternative fuels. Here the present-day cost of providing 1 therm (105.5 MJ) of heat is shown, related to the efficiency of combustion. A curve is given showing the cost of providing 1 therm of heat by means of a heat pump having values of R_η between 4 and 2. These values have been equated with a value of fuel combustion efficiency varying from $\eta = 1$ and $\eta = 0$.

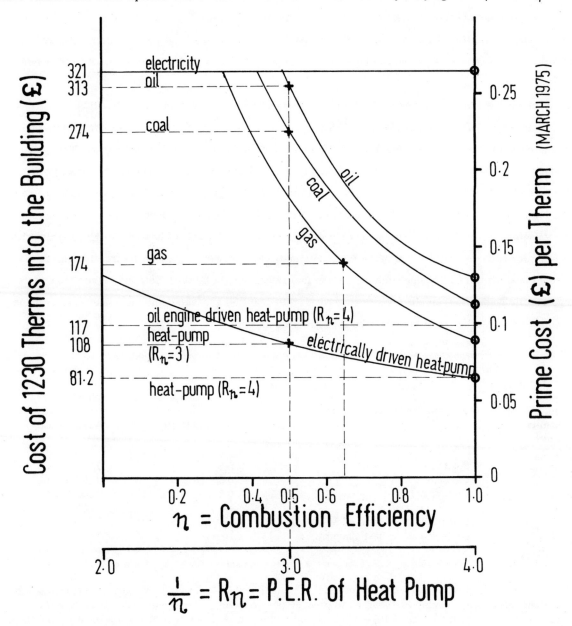

Fig. 4. Relative cost per therm of heat produced by heat pump and alternative combustion methods.

In order to obtain an assessment of the annual cost of providing 1 therm/h of heat, the relative capital costs of providing each type of fuel-burning equipment must be considered. Reasons are given later in the book which suggest that a mass-produced heat pump using either electric or oil engine drive should cost no more than fuel-burning equipment for alternative methods; capital cost of plant would then be equal for each method. The factor of annual capital cost has not, therefore, been introduced into the curves in Fig. 4.

There is only a small monetary or fuel-saving advantage in providing an electrically driven heat pump which has an averaged seasonal P.E.R. $(R_\eta) = 2$. But there is no reason why time and money should be wasted on such an inefficient machine when it is possible to manufacture heat pumps which will have a minimum seasonal P.E.R. $(R_\eta) = 3$. In the latter case both monetary and fuel saving is shown to be considerable.

A consumer requiring 1230 therms (130 GJ) who installs a heat pump with a P.E.R. $(R_\eta) = 3$ will save annually between £70 and £200 (at March 1975 prices). But let it be assumed that, after installing the machine, there was no saving in annual cost but a two-thirds reduction in fuel consumption. The question then arises as to which method of using fuel should be allowed to prevail. A world which can allow its entire store of fuel oil supplies to be consumed in a single century seems unlikely to give a rational reply.

The reader will no doubt ask at this stage what is meant by 'relative efficiencies' in various methods of fuel consumption. This value may be described as the fraction of a unit of heat usefully available after combustion for each unit of heat which the fuel contained. Only general values of the fraction can be given because so many factors influence each case when fuel is burned. In Table 1 Kell and Martin* postulate a range of efficiencies for 'small' boilers, usually larger than the average domestic boiler. Tests made myself in 1968 on twelve small domestic boilers are also shown.

Table 1. Table showing heat available for use as ratio of heat contained in fuel for various methods of combustion.

Heat in unit quantity of fuel	Method of combustion	Heat available after combustion	
		Kell and Martin	Author's tests
1.0	Solid fuel (hand-fired)	0.5	0.3 to 0.55
1.0	Gas boiler	0.65	0.55 to 0.68
1.0	Oil-fired boiler	0.5	0.38 to 0.5
1.0	Electric radiator	–	0.33†
1.0	Electrically driven heat pump (R_η=3)	–	0.99
1.0	Oil-engine driven heat pump (R_η=4)	–	1.66

† This figure assumes a generation efficiency of 38 per cent and a transmission and distribution loss of 5 per cent.

The case of the heat pump driven by an oil engine is analysed later in the book (see Appendix B). The power output from the oil engine is assumed to provide power having a heat equivalent of 0.33 of the heat available in the fuel. With a value of $R_\eta = 4$ this would make available as heat at t_1, $4 \times 0.33 = 1.32$ times the heat in the fuel.

* Kell and Martin. Heating and Air Conditions of Buildings, Architectural Press, 1968.

4: How heat pumps work

Heat Pumps in Operation

An examination will now be made of the practical operation of typical heat pumps, using air as the 'working substance', i.e. the refrigerant. The thermodynamic principles were explained in Chapter 3 and a more detailed description is contained in Appendix A.

The domestic heat pump is primarily a refrigerator but contrary to general opinion it is not a 'reversed' refrigerator, although it is a 'reversed' heat engine in the sense that it has to be driven by an external source of power. The first function of any heat pump is to refrigerate, i.e. to extract heat from its surroundings and to produce and maintain, during this first function, heat at temperatures below that of the surrounding atmosphere. Its final prime function is to lift the temperature of the heat extracted by its refrigerator to a higher temperature, by compression, for use in providing heating. It will be seen that a refrigerator is used primarily to make things colder than the surrounding atmosphere, whereas the heat pump refrigerates as a secondary process with the primary function of producing heat at temperatures well above that of the surrounding atmosphere.

We will deal now with the secondary function of refrigeration which may, of course, be produced in a number of ways. In this book only two ways are considered, namely, the expansion and compression of air in such a manner that the air temperature falls during expansion below the surrounding air temperature, and secondly, the principle of allowing the expansion of a fluid and its vapour to take place so that the change in state can extract heat from the atmosphere and then, by subsequent further heating due to compression of the vapour, convert that heat to a higher temperature. It is this second method which is the main feature of this book.

In Fig. 5a a simplified diagram is given for an expansion and compression machine using air as a refrigerant. M is a direct-coupled driving motor. Air from receiver X is compressed (operation bc in Fig. 5b) to a high temperature t_1' into the heat exchanger (receiver Y) when it gives up the heat to the room (operation cd). The air, therefore, enters the expansion engine E at some temperature t_1 lower than t_1' but still retaining some portion of the pressure imparted to it by the compressor (where work w_c was put in). It therefore expands in E and does work w_e (operation da) such that the net work done in the cycle is w_n and is equal to $w_c - w_e$, and in this expansion, because work is done, the air temperature falls below the atmospheric temperature t_2 to a lower temperature t_3. It is, therefore, taken into the second heat-exchanger, receiver X (operation ab) where the air temperature can rise to that of the atmosphere.

This cycle is referred to by writers in Britain as the 'reversed Joule cycle' and in the USA as the 'reversed Brayton cycle'.

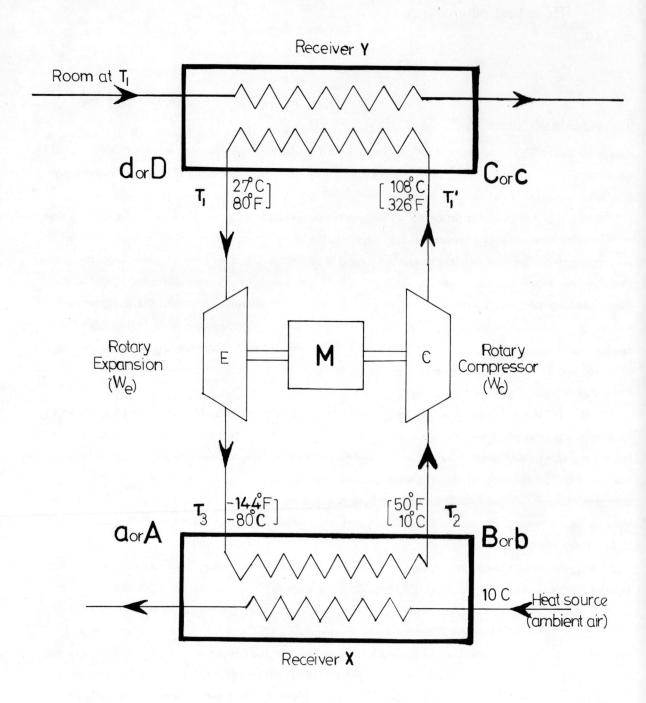

Fig. 5a. Mechanism of rotary type of refrigerator/heat pump using air as working substance and working on Joule reversed cycle.

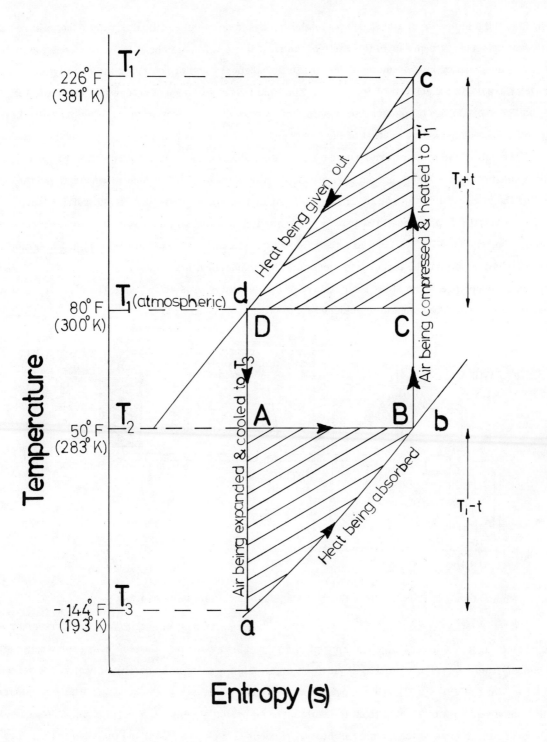

Fig. 5b. Joule cycle operation (abcd) compared with Carnot cycle (ABCD).

In Fig. 6 the elements of a heat pump are shown, based on the use of liquid, rather than air, as the refrigerating agent. It represents the parts of the first large heat pump built in Britain (see Fig. 28), which was designed by myself and made by my staff (except the compressor) for the main building of the Norwich Corporation Electricity Department. The liquid refrigerant must be one (of many available) which will 'boil' and evaporate at low pressures with their associated low temperatures. The pressure in the evaporator is reduced by the suction of the compressor to around 35 p.s.i.a. River water is circulated through the coil in the evaporator at river temperature, thus causing the the refrigerant to 'boil'. In this state, changing from a liquid to a vapour, a considerable amount of latent heat is required and this is provided by the river water passing through the coil. At this stage, about 10 lb of liquid refrigerant, passing through a normal 5 h.p. domestic heat pump each minute, has turned into a vapour at or near river water temperature, say 40°F (4.4°C) and has abstracted approximately 600 Btu/min (176 W) of <u>low-grade</u> heat from the river supplying the circulating medium. This vapourised low-grade heat at 40°F (4.4°C) is now drawn into the compressor and the temperature of the vapour is boosted to 120°F (48.9°C) or whatever higher temperatures the machine is designed for.

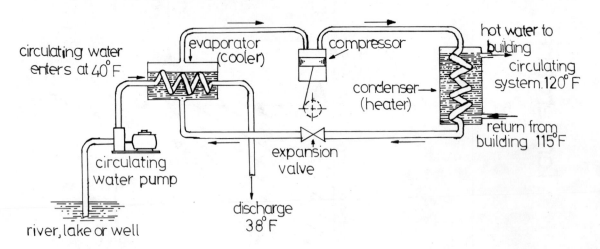

Fig. 6. Elements of the Norwich Heat Pump, using vapour compression cycle.

At this stage of the operation, therefore, we have taken an amount of heat (q_2) from the river or other source at 40°F (4.4°C), then added an amount of compression work (w_c) thereby lifting the temperature of the vapour from 40°F (4.4°C) to 120°F (48.9°C) or t_1. This results in a quantity of heat in the refrigerant vapour (q_1) at 120°F (48.9°C). This vapour now enters the condenser heat-exchanger where it is surrounded by cooled warm water coming from the building to be heated at, say, 115°F (46.1°C). Consequently the vapour containing some of the quantity of heat q_1 condenses back to a liquid and in so doing transfers its latent heat q_1 to the building water passing through the condenser unit. As in the air-type heat pump of Figs. 3c and 5a, we have at this stage a refrigerant which has been compressed to a high pressure. In the air-type machine it was possible to let the higher-pressure air do work by expansion and so to reduce the total amount of work expended to $w_n = w_c - w_e$. There is no technical reason why this should not be done in the case of the liquid refrigerant, using a small turbine attached to the compressor driving shaft. It has, however, not been usual practice and it is normal for the pressure and temperature of the liquid in the condenser to be reduced by means of the expansion valve shown in Fig. 6. so that the refrigerant now re-enters the evaporator to commence a new cycle.

Ideal Cycle of Operation

It is necessary, now, to ascertain whether a machine can be made which will enable the four operations to be carried out as shown in cycle ABCD (Fig. 3d) which represents an <u>ideal</u> cycle termed the 'reverse Carnot cycle'. In fact, a machine following the Carnot cycle is impracticable for these reasons:

(a)   AB and CD represent 'isothermal' operations in which heat is taken in from, say, soil and given out, say, to heated circulating water without any temperature difference or change. With air, in particular, there must be a temperature difference if a heat exchange is to occur.

(b)   The compression and expansion (BC and DA) must be 'adiabatic', i.e. no heat must be lost from the gas to the pistons, cylinder wall etc. or by friction. This is not possible in practice.

Because no heat engine or heat pump can operate on this perfect cycle where

$$R_{\eta\,(I)} = \frac{t_1}{t_1 - t_2} \,,$$

it will be used as a standard by which a heat pump can be judged.

Practical Cycles of Operation

There are several <u>practical</u> cycles of operation which can be used to operate a heat pump. Each of them is the 'reversed' cycle of some form of driving engine which takes in heat and utilises some part of the heat taken to produce power and the ability to do work. Examples are the steam engine and steam turbine and the various types of hot-air engines invented by Kelvin, Joule and Stirling early in the nineteenth century. If the operation of any of these heat engines <u>doing work</u> to drive a motor is being considered as a cycle, it would be necessary to commence at point D in the indicator diagram in Fig. 3d, and to proceed <u>clockwise</u> around the cycle, i.e. DC to CB to BA to AD. Since, however, we are now going to drive the heat pump from a motor, we shall be following the 'reversed cycle' already considered, i.e. AB to BC to CD to DA in an <u>anti-clockwise</u> direction. The following two cycles will be considered:

1.   Joule reversed cycle, using air as the refrigerant (Figs. 5a and 5b).

2.   A modified Rankine reversed cycle, using a liquid and its vapour as the refrigerant (the direct Rankine cycle is used in the steam engine) as shown in Figs. 6 and 7a. This will subsequently be termed a 'vapour compression' cycle.

It will be shown later that even these practical cycles cannot be achieved ideally and that approximately 60 per cent of the ideal Carnot operation will be achieved in practice.

There are two regressions in efficiency from the ideal Carnot reversed cycle when operating a heat pump. The first regression is caused by the need to have a temperature difference in both condenser and evaporator before there can be a heat exchange between vapour and water or air, i.e. the temperature change is not isothermal at t_1 and t_2 but occurs at $t_1 + t$ and $t_2 - t$. The second regression occurs because, during both compression and expansion, heat escapes via cylinder walls and piston and is no longer an ideal adiabatic process.

As an example: a Carnot-cycle heat pump with R_η (Carnot = $R_{\eta(I)}$) could work between temperatures of 40°F (4.4°C) and 140°F (60°C) and have an 'efficiency' or Performance Ratio (R_η) of:

$$R_{\eta(I)} \quad \frac{t_1}{t_1 - t_2} \quad = \quad \frac{333^\circ K}{333^\circ K - 277^\circ K} \quad = \ 6.0 \ .$$

But, in a practical machine, with $R_{\eta(P)}$, there would be a 10°F (5.5°C) drop in the evaporator and condenser respectively in order to get heat transfer. Due to this, the machine must work between 30°F (-1.1°C) and 150°F (66°C):

$$R_{\eta(P_1)} \quad \frac{t_1 + t}{t_1 - (t_2 + 2t)} \quad = \quad \frac{339^\circ K}{339^\circ K - 272^\circ K} \quad = \ 5.0 \ .$$

So that the temperature drops alone which are necessary in a practical machine have reduced the efficiency by 16.7 per cent below the ideal (Carnot) efficiency having 'isothermal' heat exchange. It was also shown above that $R_{\eta(I)} = \dfrac{t_1}{\text{work done}}$ and assumed that work was done 'adiabatically', i.e. without any heat loss to piston or cylinder walls. In practice this heat loss may represent an additional 40 per cent of work/heat giving a final practical value very approximately as follows:

$$R_{\eta(P_2)} \quad = \quad \frac{339^\circ K}{(339^\circ K - 272^\circ K)} + 40\% = \frac{339^\circ K}{94^\circ K} = \ 3.6$$

when $\quad R_{\eta(P_2)} \quad = \ \dfrac{3.6}{6.0} \ = \ 60\% \times R_{\eta(I)}.$

EXAMPLES SHOWING CALCULATION OF CYCLE VALUES

Since the more serious reader may wish to follow in detail the design of compressor and components as shown in the design project in chapter 7 it will be necessary to become familiar with the estimation of heat quantities taken from pressure/enthalpy (p/h) charts of the types shown in Figs. 7a, etc.

Two examples of ideal operation, one using air and the other liquid/vapour as a refrigerant, are now given and the method of calculating heat quantities is shown.

EXAMPLE 1 - Air Cycle

A heat pump using 1 lb (0.454 kg) of air per minute is to be designed to operate on the Carnot cycle ABCD of Fig. 3d. The air must enter the compressor at point B when it is at a temperature of $50^{\circ}F$ ($283^{\circ}K$) and be compressed to point C (adiabatically), i.e. at constant entropy, until it reaches a temperature of $80^{\circ}F$ ($300^{\circ}K$) at point C. It must then give up its heat 'isothermally', i.e. without falling in temperature, after which it enters the expansion engine at point D, where it expands and does work isentropically during operation DA, during which it falls to its original temperature of $50^{\circ}F$ ($283^{\circ}K$) at point A. The air must then take in heat at constant temperature until it has increased in volume and decreased in pressure so as to return it to its original condition at point A, when the cycle can recommence.

If this cycle was practicable, we could say that:

$$\text{P.E.R.} = R\eta_{(c)} = \frac{\text{Heat taken in}}{\text{Net work done}} = \frac{t_1}{t_1 - t_2}$$

and

$$R\eta_{(c)} = \frac{300^{\circ}K}{300^{\circ}K - 283^{\circ}K} = \frac{18}{1}. \quad \text{(See Appendix A)}$$

The above (Carnot) cycle is not practicable, because the air entering receiver R at $300^{\circ}K$ and receiver Y (Fig. 5a) at $283^{\circ}K$ could not give up or take in heat when the two elements in each receiver were at the same temperatures. In practice, therefore, we must use a heat pump working on the reversed Joule cycle as in Fig. 5b and the temperature at C in this figure must be increased to a higher temperature at point C and the temperature after expansion must fall to point a instead of point A. In practice the new temperatures for the reversed Joule cycle abcd in Fig. 5b might be:

Point B (or b) $t_2 = 50^{\circ}F = 283^{\circ}K$

Point c $t'_1 = 226^{\circ}F = 381^{\circ}K$

Point D (or d) $t_1 = 80^{\circ}F = 300^{\circ}K$

Point a $t_3 = -144^{\circ}F = 193^{\circ}K$.

For this more practical air cycle:

$$\text{P.E.R.} = R\eta_{(P)} = \frac{381^{\circ}K}{381^{\circ}K - 283^{\circ}K} = 3.88 = \frac{t'_1}{t'_1 - t_2}.$$

It will be noted how the need to have temperature differences and therefore a higher value of t_1 and a lower value of t_2 has reduced the value of $R\eta_{(P)}$ for the reversed Joule cycle to 21 per cent of $R\eta_{(I)}$ for the reversed Carnot cycle; values in the example are exaggerated.

This is clearly indicated in Fig. 5b. The work done in the reversed Joule cycle is represented by area abcd and that in the Carnot cycle by area ABCD, the former being greater in value than the latter for a somewhat similar heat output.

Now consider the 'vapour compression' cycle shown in Figs. 7a and 7b. It will be noted that the regression of the reversed Joule cycle in Example 1 is due to the increased area of cCD and

BaA shown in Fig. 5b. To reduce the effect of this regression, we require to replace the constant pressure lines cd and ab of Fig. 5b (during which heat was given out and taken in) by a cycle in which heat can be given out and taken in at as near as possible to 'constant temperature', i.e. isothermally along lines CD and AB in Figs. 5b, 7a and 7b. This may be approached if we use a vapour instead of air as a refrigerant. By using a vapour the operation AB in Fig. 7a would represent the evaporation of a liquid and the line CD the condensation of the vapour. We require a liquid that will boil (evaporate or condense without changing temperature) at low pressures and temperatures. A homely example is the liquid methane, which can be purchased in tubes for the purpose of filling cigarette lighters. If a small amount of this liquid is reduced in pressure by being released on the hand, it will boil vigorously at atmospheric pressure without changing temperature. In thus changing from a liquid to a vapour, it will take in a large amount of 'latent heat' from the skin, and in sufficient quantity, can cool the skin so as to cause a 'burn'. In Figs. 7a and 7b heat is absorbed (q_2) from air or water as the liquid 'boils' and turns into a vapour along line AB, i.e. isothermally. Then the vapour is compressed (w_c) along line BC and heated from t_2 to t_1. Finally, the heat equivalent of $q_2 + w_c$ contained in the refrigerant at t_1 (still in vapour form) is condensed along line CD. Fig. 7a represents the above vapour compression cycle when drawn between ordinates of pressure (vertically) and total heat of the refrigerant (horizontally). The type of p/h diagram shown in Fig. 7a will be used throughout the remainder of the book, since we shall be studying subsequently only the vapour compression cycle.

EXAMPLE 2 – Vapour Compression Cycle

We will now consider a heat pump working between the same temperatures as for Example 1 but using a liquid and its vapour as the refrigerant when operating on the ideal vapour compression cycle shown in Figs. 7a and 7b with t_1 and t_2 as in Example 1:

$$t_1 = 80^{\circ}F \ (26.7^{\circ}C) \qquad p_1 \ (\text{refrigerant } CF_2Cl_2) = 98.76 \text{ p.s.i.a.} (6.72 \text{ bar})$$
$$t_2 = 50^{\circ}F \ (10^{\circ}C) \qquad p_2 \ (\text{refrigerant } CF_2Cl_2) = 61.4 \text{ p.s.i.a.} (4.18 \text{ bar})$$

To solve Example 2, first draw a horizontal line AB as in Fig. 7a between LL and VV at 61.4 p.s.i.a., which is the refrigerant pressure at t_2, followed by a second horizontal line at CD at 98.76 p.s.i.a., representing t_1 at $80^{\circ}F$. Then drop a vertical line from where the upper horizontal intersects line LL at D (with value $H_D = 26.28$ Btu/lb (7.7 W/lb)) to the state point A on the lower horizontal line, giving state points D and A with equal values of H. Mark state point C at the intersection of DC with line VV (having value $H_C = 86.8$ Btu/lb) as it will be assumed here that the refrigerant will be dry saturated after adiabatic compression between B and C. By following the 'constant' entropy curves we can draw line CB from C to B to give state point B with a value of $H_B = 82.82$ Btu/lb. Then w_c which equals $H_C - H_B$ has a calculated value of 3.98 Btu/lb (1.17 W/lb). The cycle diagram of Fig. 7a is now completed and values of heat units can be read off by the projection of state points A, B, C and D down to the base line. Values for the cycle are as follows:

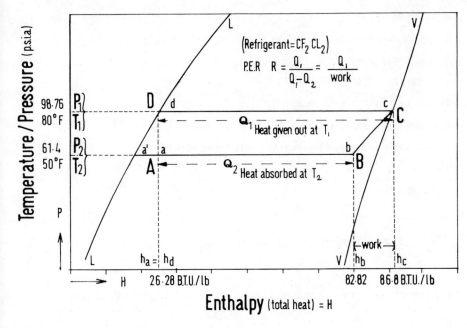

Fig. 7a. Pressure/enthalpy (p/h) cycle diagram for cycle in Example 2.

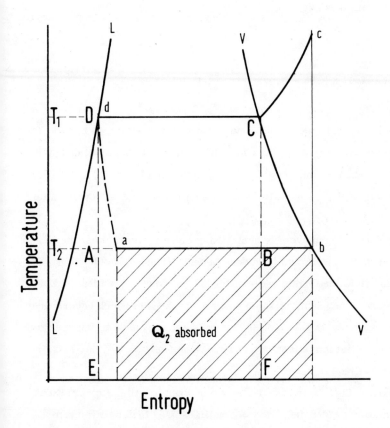

Fig. 7b. Temperature/entropy (TS) cycle diagram for cycle in Example 2.

<u>1st operation</u> : line AB = heat taken in at $t_2 = q_2 = (H_B - H_A)$

$\qquad$ = (82.82 - 26.28) = 56.54 Btu/lb (16.6 W).

<u>2nd operation</u> : line BC = adiabatic compression = $w_c = (H_C - H_B)$

$\qquad$ = (86.8 - 82.82) = 3.98 Btu/lb (1.17 W).

<u>3rd operation</u> : line CD = heat given up at $t_1 = q_1 = (H_C - H_D)$

$\qquad$ = (86.8 - 26.28) = 60.52 Btu/lb (17.74 W).

<u>4th operation</u> : line Da = the condensate is passed through the expansion valve at constant heat so that

$$H_D = H_A.$$

From the above we have:

$$\frac{q_1}{q_1 - q_2} = \frac{60.52 \text{ Btu/min}}{60.52 - 56.54} = 15.2 = \text{Performance Ratio} = R_{\eta_{(I)}}.$$

Reference to chapter 7 will show how further properties of the cycle in Fig. 7a may be calculated:

<u>Quantity of refrigerant passing</u> = q_r

$$q_r = \frac{H_c - H_d}{H_b - H_a} = \frac{60.52}{56.54} = 1.07 \text{ lb/min (0.46 kg/min)}.$$

<u>Indicated horsepower</u> = I.H.P.

$$\text{I.H.P.} = q_r \times (H_C - H_B) \times 0.0236 = 1.07 \times 3.98 \times 0.0236 = 0.1 \text{ I.H.P.}$$

Therefore, an <u>ideal</u> heat pump working on the vapour compression cycle between temperatures t_1 and t_2 as specified in Example 2 would, for a heat output at t_1 of 60.52 x 1.07 Btu/min (20.44 W) or 4185 Btu/hr (1226 W), require a work input of 0.1 I.H.P. (82.5 W) and would have a Performance Ratio $(R_{\eta_{(I)}})$ of 15.2. This value can be compared with a Carnot value $(R_{\eta_{(c)}})$ equal to 18 and with a reversed Joule cycle value of 3.88 for machines working between $80°F$ ($26.6°C$) and $50°F$ ($10°C$).

The value of $R_{\eta_{(I)}}$ for the ideal vapour compression cycle is remarkably close to the value of the Carnot cycle. In fact, except for the fact that no expansion work has been done between D and A, the ideal vapour compression cycle as shown in Figs. 7a and 7b <u>is</u> the Carnot cycle. This fall of value in the ideal P.E.R. from R_η = 18 to R_η = 15.2, or approximately 10 per cent, is some measure of the loss which occurs when the pressure and temperature change from state points D to A in the cycle is accomplished by using an expansion valve instead of an expansion engine. Alternative methods of retrieving some part of this loss, by sub-cooling and other means are dealt with in Chapter 8.

The ideal cycle shown in the p/h diagram of Fig. 7a and the T/S cycle of Fig. 7b will suffer practical losses that will probably halve the ideal value. The T/S diagram is sometimes used to delineate cycle ABCD because it gives a value of heat which is directly proportional to area.

Review of Heat Pump Cycles

The first heat pump using air as the refrigerant was proposed By W. Thomson (Lord Kelvin) in 1852. The cycle was later, probably wrongly, referred to in Britain as the reversed cycle of Joule's hot-air engine. Kelvin states that 'The method of cooling air...has been realised by Mr. Coleman...in the Bell-Coleman refrigerating machine'. In the USA the same cycle is described as the reversed Brayton cycle.

The advantages given by using a vapourisable liquid as the refrigerant were soon realised (Kelvin refers to its use in his article) and a modification of the reversed Rankine or 'vapour compression cycle' using a vapourisable liquid ultimately replaced air as the refrigerant. This cycle gave a much closer approach to the isothermal operations of taking in and giving out of heat as postulated in the ideal Carnot cycle; better values of P.E.R. were also obtained.

Air used as the working substance, i.e. refrigerant medium, such as in the Rotary Pressure Exchanger suggested by Lebre, may have potentialities capable of future development. The use of air as the refrigerant led to difficulties in operation and in design and the advantages of a vaporisable liquid instead of air as the working substance led to a fairly general adoption of the vapour compression (reversed Rankine) cycle. Machines using air were bulky and resulted in a large waste of power through friction. The use of a liquid with a high latent heat of vaporisation as the working substance for a heat pump in a more compact and efficient machine also permits of a much higher P.E.R. being achieved than with air. Hence closer approach can be made to the Carnot cycle. This is shown in Table 2 which indicates the performance for a heat pump, using various working substances, ideal reversed cycle values being used throughout.

Table 2. Ideal cycle values of R_η for various refrigerants when $T_1 = 110°F\ (43°C)$ and $T_2 = 35°F\ (1.6°C)$.

Working substances	Unit	Ideal cycle values Ideal Carnot	Free air	Water	F22	F12	F114	F11
P_2	lb/in² abs.	—	14.7	0.1	76.6	47.3	13.7	6.2
P_1	lb/in² abs.	—	73.5	1.275	243	151	54.4	28.1
P_1/P_2	—	—	5.0	12.75	3.19	3.19	4.0	4.53
Refrigerating effect	Btu/lb	—	32.4	999	64	48.5	41.5	66.1
Heat delivered as heat pump	lb	—	51	1192	76.3	57.7	49	77.4
Ideal compressor work	lb	—	18.5	271	12.3	9.14	7.5	11.3
H.p. input (as refrigerator)	h.p./ton	—	2.69	1.28	0.908	0.88	0.853	0.808
H.p. input (as heat pump)	h.p./1000 Btu/min	—	8.57	5.4	3.8	3.74	3.61	3.44
C.O.P. (refrigerator)	C.O.P.	6.6	1.75	3.40	5.2	5.3	5.53	5.85
P.E.R. (heat pump)	P.E.R.	7.6	2.85	4.4	6.2	6.3	6.53	6.85
Efficiency (as refrigerator)	$R_{\eta(I)}$	1.0	0.265	0.56	0.788	0.805	0.838	0.885
Efficiency (as heat pump)	$R_{\eta(P)}$	1.0	0.362	0.58	0.816	0.829	0.859	0.9

A curious feature should be mentioned here. The use of air as the working substance in an air-cycle machine such as that shown in Fig. 5a was superseded in refrigeration and heat pump work by the vapour compression cycle. But recently, air-cycle heat pumps for heating and/or cooling have become general for use in aircraft. Fig. 8 is a diagram of an aircraft heat pump consisting of a combined supercharger and pressuriser for air-comfort and cooling use.

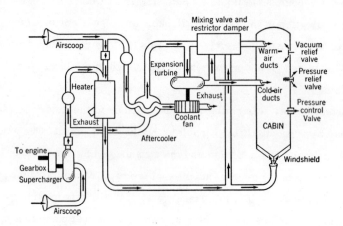

Fig. 8. Air heat pump as used in modern aircraft. (Reproduced from Heating, Piping Air Conditioning.)

5: Capturing low-grade heat

General Observations (see Appendix C).

In Chapter 3 it was mentioned that we can convert all of a given amount of work into heat but that only some fraction of heat commencing as a quantity q_1 can be converted into work. We showed that if fuel was burned in a heat engine, we should obtain an amount of heat equal to q_1 at t_1; the process gave (assuming $q_1 = 3$):

$$3 \;=\; 1 \;+\; 2$$

Heat in fuel (q_1) = Work + Rejected Heat (q_2), i.e. $\dfrac{\text{Work}}{q_1} = \dfrac{1}{1+2} = \dfrac{1}{3}$. (1)

Next, means were considered which would permit us to reverse the process by using a heat pump so that:

$$2 \quad + \quad 1 \quad = \quad 3$$

Rejected Heat (q_2) at t_2 + Work (w) = Heat to room (q_1) at t_1, i.e. $\dfrac{q_1}{w} = \dfrac{2+1}{1} = \dfrac{3}{1}$. (2)

In this chapter we must quantify the likely values of q_2 and w. Beneficent Nature's Law seems to be that we can have all that we require if we are willing to work to obtain it – whether it be harvest after seedtime or heat after wasting our substance, it will always be there for our use. We shall, therefore, look at our arb–itrary reversed cycle equation again and derive some quantitative values for q_2 as related to the other two factors, i.e. w and q_1. Values obtained, in practice, for a given case, might be as follows:

$$q_2 \;(\text{at } t_2) \qquad + \quad w \;=\; q_1 \;(\text{at } t_1), \text{ i.e.}$$

Unavailable Heat	+	Work	=	Available Heat		(3)
83%	+	17%	=	100%		

or, say,

$$38,000 \text{ Btu/h} + 7600 \text{ Btu/h} = 45,600 \text{ Btu/h} \qquad (4)$$
$$(11,140 \text{ W}) \qquad (2280 \text{ W}) \qquad (13,420 \text{ W})$$

This chapter, therefore, examines the various ways in which large quantities of semi-hot, low-grade heat (which may be four or five times the work/heat input) at, or near, ambient temperatures can be collected and used for the purpose of being up-graded to heat of a more useful nature.

The Economic Value of Low-Grade Heat

Each unit of low-grade heat which can be collected has a potential monetary value at least as great as each heat unit finally obtained from coal or oil. This can be shown by assuming that 1230 therms (116.6 GJ) of heat at t_1 are required to heat a house during the heating season. If oil were burned at 50 per cent combustion efficiency, 6.13 tonne would be consumed. If an oil engine were used to drive a heat pump ($R_\eta = 4$), the heat pump would deliver 1230 therms with an oil consumption of only 2.25 tonne ($\eta = 0.33$ for the oil engine), thereby saving 3.9 tonne of oil. The standard equation will have the following values:

$$q_1 \qquad = \text{Work} \qquad + \qquad q_2$$

$$1230 \text{ therms at } t_1 = 308 \text{ therms} \quad + \ 922 \text{ therms at } t_2.$$

(Heat to house)   (from oil engine)   (from low-grade heat)

3.9 tonne of oil per year will have been saved worth £240 by up-grading 922 therms of low-grade heat; this gives a value of 1 therm of low-grade heat as being £0.26 (see Appendix C).

Practical Values of Low-Grade Heat

We may use the expressions referred to earlier to find the amount of work/energy that must be provided to lift a given quantity q_2 of Unavailable heat taken from our surroundings at t_2 so as to provide Available heat at t_1. Assume that q_1 represents the heat energy equivalent to one electrical unit (1 kWh or 3412 Btu/h) and that the low-grade heat is available at 40°F (4.4°C). Table 3 shows the amount of low-grade heat and the necessary power input, respectively, to provide 1 kWh of heat at temperatures t_1 which vary from 60°F to 200°F (15°C to 93.3°C).

Table 3. *Quantities of low-grade heat (Q_2 at 40°F) and power input W to provide the heat equivalent of 1 kWh at various temperatures = T_1*

Temperature T_1 (°F)	60	80	100	120	140	160	180	200
(°K)	288.6	299.7	310.8	321.9	333	344.1	355.2	366.3
Q_2 (kWh)	0.96	0.90	0.88	0.86	0.85	0.82	0.80	0.78
W (kWh) Ideal	0.04	0.07	0.11	0.14	0.17	0.19	0.22	0.24
W (kWh) Practical	0.08	0.14	0.22	0.28	0.34	0.38	0.44	0.48

The practical motor input, allowing for losses, is based on a value of $R_{\eta(P)} = 0.5\,R_{\eta(I)}$.

Table 3 clearly shows the advantage of keeping the upper temperature t_1 as low as possible; e.g at a temperature of 200°F the work input is approximately twice that required if t_1 were maintained at 110°F.

Table 4.   Table showing low-grade heat collection by various types of coils using F12 (CF_2 Cl_2) as refrigerant.

Evaporator Pressure Temp:	Anti-Freeze In	Out	Refrigerant Flow Per Minute	Condenser Pressure Temp:	Motor Load (W) K.W.	Coil in Use
p.s.i.a.	°F		lbs			
26	33	44	11.8	144	4.2	A + B*
22	31	48	11.1	148	4.1	A only
22	22	33	10.2	150	4.0	B only*
18.5	20	33	8.5	140	3.88	B**

Calculated Values of Q_1 and Q_2 etc.

Q_1 Per Hour B.Th.U	Watts	Q_2 Per Hour B.Th.U	Watts	Heat Collected by Coil = Q_2 Per Hour Per Lineal Foot B.Th.U	Watts	Lineal Metre B.Th.U	Watts	P.E.R.	Coil in use
44140	12940	35952	10540	58.94	17.28	17.98	5.27	3.0	A*+ B
40860	11980	32912	9650	219.4	64.2	66.02	19.61	2.9	A*
31020	9836	23120	6720	50.2	16.78	17.47	5.12	2.46	B*
30260	8860	22480	6588	48.85	13.8	11.22	3.29	2.3	B**
12680	3716	7100	209	35.5	19.4	10.74	3.15	—	C

Details of Coils

Coil	Length Feet	Metres	Surface Area Feet²	Metres²
A	150	45.75	29.6	2.752
B	460	140.3	92.2	8.58
A & B	610	186	121.2	11.25
C (Black Plastic Hose)	200	60.2	28.4	2.68

B* Test taken at comencement of heating season (Sept: 1974)
B** " " " " " " " (May: 1975)
A* Coil A immersed in water maintained at 47°F (8.3°C)

Table 5.   Table showing relative values of Q_2 and Ideal and Practical values for W_c when Q_1 = 36 000 Btu/h at various temperatures; Q_1 = 36 000 Btu/h = 10.55 kW = 14.07 B.H.P.

T_1 (°K)		288.6	299.7	310.8	321.9	333	344.1	355.2	366.3
T_1 (°F)		60	80	100	120	140	160	180	200
Q_2 (Btu/h)		33 230	30 580	28 170	26 000	24 000	22 000	20 300	18 900
W_c (Btu/h)		2 770	5 420	7 830	10 000	12 000	13 900	15 700	17 000
W_c (kWh)	Ideal	0.4	0.8	1.1	1.7	1.75	2.0	2.3	2.5
W_c	Practical	0.81	1.6	2.2	3.3	3.5	4.0	4.6	5.0
W_c	B.H.P.	1.08	2.2	2.9	4.4	4.7	5.3	6.1	6.7

This table shows that we can have either a low value of low-grade heat input Q_2 with a high work input W_c or a lower work input and a necessarily higher value of Q_2.

Terrestrial Heat

The earth possesses large quantities of ambient heat derived from several sources. The air collects the heat rejected from the combustion of fuel and may be very rich in heat. Air temperatures taken in a street im the centre of London have been shown to be as much as 8 deg C higher than temperatures recorded at the same time sixteen miles away.

Heat from the sun penetrates the earth slowly by conduction from the surface downwards and the maximum temperatures are therefore at different depths at different times. An experiment was carried out in Edinburgh in which four thermometers were sunk at depths of 1, 2, 4 and 8m into rock; the maximum temperatures were recorded in August, September and October and it was found that some of the heat was retained in each layer which raised the temperatures so that yearly variations diminished with depth. The temperatures were, respectively, $8.3°C$, $5.6°C$, $2.7°C$ and $1°C$. It was concluded that solar heat does not penetrate beyond a certain depth.

It would seem, therefore, that our problem is one of collecting solar heat stored in the earth's surface, the earth's water and the air surrounding us or, alternatively, solar heat received directly. It is a strange commentary that we willingly spend millions of pounds in researching and making machinery that will extract heat from the earth in the form of the rapidly diminishing reserves of coal or oil, to be used at an average efficiency of 50 per cent or less, and literally nothing on research and making machinery that can collect the much more valuable low-grade heat stored in earth, air and water which, if used properly, can reduce fuel consumption by up to two-thirds. Technical literature and achievement in the first method is abundant; literature and achievement in the second is scarce.

Low-Grade Heat Sources

It is proposed to consider now the various sources from which the quantity of low-grade heat (q_2 at t_2) may be obtained. We require a source which has a high specific heat per unit of volume so that the temperature of the source falls as little as possible during the abstraction of heat. As an example, for a fall in temperature of 1 deg F (0.56 deg C) one pound of air (0.454 kg) will yield 0.24 Btu and one pound of soil or water about 1.2 Btu and 1 Btu respectively, and since the natural sources of low-grade heat are earth, air and water, these three sources will be studied in that order.

Heat From Coils Laid in Soil

So far as I know, no adequate theory has been put forward as to the exact mechanism of heat transfer through the soil. It is possible that heat is transferred by conduction and by convection and radiation in the air cavities in the soil. An alternative theory is that transference is by conduction due to variation in partial vapour pressures.

The coil shown in Figs. 9 and 10 is at the centre of a cube of earth weighing 600 tonne. The heat available in this volume of earth, between 7°C and 0°C is approximately 17×10^6 Btu (17 GJ). The annual demand of my heat pump for low-grade heat is $3000h \times 28,000$ Btu, i.e. 84×10^6 Btu (84 GJ).

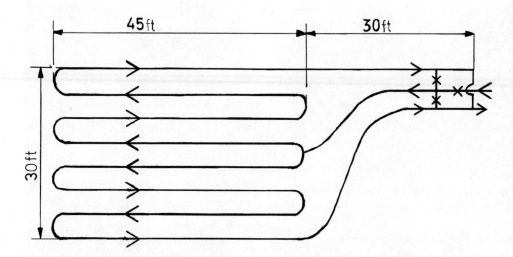

Fig. 9. Layout of buried ground coil for heat pump at a depth of 3 ft using $\frac{3}{4}$ in bore soft copper pipe.

Theoretical calculations as to coil lengths, using standard heat transfer equations, give a guide as to the lengths of buried coil required. The value of length for the coil shown in Figs. 9 and 10, determined from the steady-state equation would be L= 1256 ft (382 m) and $q_{/L}$ = 23.9 Btu/h/lin ft. But investigation showed that the buried coil began to form a covering of ice at the beginning of each season and that this coating increased in diameter as the season progressed. Therefore, we now have to consider the effect when we have three radii to consider, i.e. pipe, ice and earth. In this case the steady-state equation gives values not appreciably different, i.e. L = 1260 ft (350 m) and $q_{/L}$ = 23.8 Btu/h/lin ft.

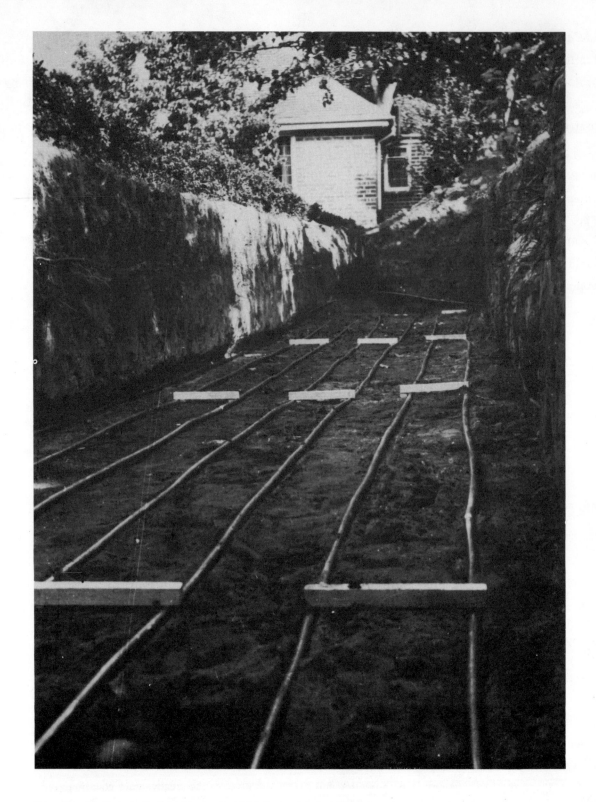

Fig. 10. Photograph of coil being laid at the author's house.

42

In fact, during the first short period of operating the earth coil, the heat demand from the coil usually exceeds the rate at which heat can be transmitted through the soil. As a result ice begins to form on the metal surface of the coil. The above calculations suggest that the conductivity of the ice is not seriously less than for the surrounding soil. When, subsequently, the heat pump stops and the ice begins to melt, the latent heat of fusion becomes stored for future use. Cessation of operation results in some part of the ice being melted by absorbing heat from the higher-temperature soil, thus giving the storage of a large quantity of heat adjacent to the iced coil surface.

The fact that, by the end of the heating season, the diameter of the pipe increases due to the ice raises the question of the possible loss of efficiency and the reduction in value of R_η due to the fairly large temperature drop between the antifreeze circuit and the refrigerant (some part of which is inherent in heat exchanger design). Because of the ice formation the antifreeze circuit maximum temperature cannot rise above 32°F $(0^\circ$C). The corresponding temperature of the refrigerant at 35 p.s.i.a. (2.4 bar) in the evaporator is 19°F $(-7.2^\circ$C) giving a temperature fall of 13 deg F (7.2 deg C) which is excessive. It is possible to reduce this loss of efficiency by introducing the refrigerant directly into the ground. It must be noted, however, that there are certain hazards if the ground coil is used as the evaporator with direct circulation of the refrigerant.

The first danger that may arise if the refrigerant is introduced directly into the ground coil concerns the question of oil return. There is the possibility that entrained oil might become trapped. This could reduce the heat transfer capacity but, more seriously, might leave the crank case short of oil. In addition, there is the danger of vapour locks forming at coil bends. The coil must be at such a height and slope above the compressor as to ensure oil and refrigerant return but not so high as to cause excessive pressure drop at suction. Experiments made some years ago in the USA using direct refrigeration circulation led to the suggestion that the length of coil should approximate to 200 ft (61 m) of 0.875 in (2.2 cm) bore for each nominal horsepower of compressor activity; this would suggest a coil length of 1100 ft for the 5.5 B.H.P. machine being considered later.

Research on buried pipe coil conductivity carried out by Miss. M. Griffith for the B.E.I.R.A. Heat Pump Committee led to the conclusion that in London clay, at depths up to 6 ft (1.85 m), the steady rate of heat absorption varies from 30 - 60 Btu/h/lin ft (8.8 - 17.6 W/lin ft). These tests used $\frac{3}{4}$ in (1.9 cm) bore tube and no advantage was gained by increasing to 1 in (2.5 cm) bore.

It is likely that a careful study of the mechanism of heat transfer from the earth to a liquid circulating in the buried coils would result in the ascertainment of means to provide much higher transfer rates. A more intimate contact between the pipe and the soil may yield higher values of q_2. The formation of a large diameter of ice on the tube by the end of winter must, however, compress the soil and may leave the pipe without contact with the soil until ice has built up again.

The experimental results which are given later will show that the coil laid with dimensions as in Fig. 9 was of insufficient length. Lack of experience led to a limitation of 470 ft (144 m). The output of the low-

43

grade heat when used with either of the two smaller compressors used between 1950 and 1961 proved to be approximately 17,000 Btu/h (4982 W) with a suction pressure at the compressor of 32.2 p.s.i.a. (2.2 bar), i.e. 34 Btu/lin ft. This output is not necessarily the maximum output as was shown when a larger compressor was installed in 1961. The suction pressure increased to 33.2 p.s.i.a. (2.26 bar) and q_2 to 24,000 Btu/h (7034 W) or 51 Btu/lin ft/h.

Supplement to the Ground Coil

In order to increase the low-grade output, a further coil of 150 ft (51.6 m) x $\frac{3}{4}$ in (1.85 cm) bore copper piping with frame dimensions of 6 ft (1.96 m) x 3 ft (0.98 m) x 3 ft (0.98 m) was made and connected in parallel with the ground coil. Initially the supplementary coil was placed outside so as to abstract heat from ambient air. Later, the supplementary coil was placed in a sheet-steel tank of slightly greater dimensions than the coil frame, the tank being filled with water rejected from domestic and other sources. The roof of a nearby greenhouse fitted some years ago with two layers of glass acted as a solar coil feeding into the nearby tank.

It has been found that the intake of heat from ambient air through the tank surfaces is approximately equal to one-half of the low-grade heat extracted from the water by the coil, provided that the ambient temperature of the air exceeds 48°F (8.8°C).

More recently an experiment was carried out with a supplementary coil consisting of 200 ft (61 m) of $\frac{1}{2}$ in (1.27 cm) bore plastic hose laid on the ground. The reason for this experiment was that, if the copper coil iced up (as it does) soon after being put into use and ran with a permanent coating of ice, a plastic tube laid on the soil and exposed to the air might achieve an ice coating but would still provide heat to the circulating antifreeze fluid. The additional heat from this short length of plastic rubber coil was appreciable. Often, during periods when the ambient air was above 45°F (7.2°C), the ice coating was able to melt during rest periods. The results of this experiment were such as to justify the view that further experiments might well be justified with a longer length of plastic coil laid on, or immediately under, the surface of the soil. If, indeed, whatever material is used for the ground coil takes in heat during its initial work phase at a rate which is higher than the soil can transmit, so that ice results on the coil surface, the question arises whether the material of the coil, or even of burial in soil, is of any great importance. It does mean that the antifreeze circulating fluid will have a maximum temperature not exceeding 32°F (0°C); this appears to happen in any case. It could also mean that if the heat transfer from ice to antifreeze medium were lower with certain materials than with others, as is likely to be the case, there would be a somewhat lower maximum temperature value for the antifreeze entering the condenser from, say, plastic hose in place of thin copper tube.

Reference can be made here to the low-grade heat potential available in waste water from baths and sinks. Unfortunately, catchment of this waste water is usually dispersed and at ground level. There is also the need to ensure that heat extraction shall not proceed nearer than one or two degrees above freezing point. Probably, as higher-grade heat becomes more rare and expensive, architectural thought may consider this aspect of heat conservation worth more attention. Calculations made earlier have suggested that each unit of terrestrial heat has a value at least equal to each unit obtained from fossil fuels.

Use of Existing Streams and Pools

Any stream of water can be a very useful source of low-grade heat. In one case a small stream 3 ft (1 m) in width was available. The depth varied from 1 to 2 ft (0.3 to 0.6 m) with a flow rate of 3 ft/min (1 m/min). The maximum volume passing per hour was of the order of 540 cu ft (15.3 m^3). This meant that, for a drop of 1 deg F in the water temperature 36,000 Btu/h (10,550 W) was available to meet an actual requirement of 30,000 Btu/h (8790 W). A pool 6.5 ft (2 m) deep was dug into the bed of this stream to ensure that the evaporator could be placed directly into the pool at a depth of about 3 ft and thus continue to work if the pool became ice-bound. The obvious advantage of a water source of this nature is that the limitation of 32°F (0°C) of the circulating secondary fluid, or refrigerant, can rise in proportion to the water temperature, if the water flow is sufficient to prevent ice forming on the immersed vessel.

Use of Underground Water Supplies

In many cases underground water supplies are available in which case air and ground coils can be dispensed with. In my own case, water held in chalk fissures is found at a depth of 160 ft (49 m) below the house; in other areas less than a mile away water is found at about 20 - 30 ft (6 - 9 m). It is preferable that two boreholes should be drilled at a distance apart so as to return the cooled water to the chalk. Under no circumstances must this water be led straight to the evaporator as fine particles of chalk will quickly block the evaporator passages. Where such water is available it provides a splendid source of low-grade heat, usually at a temperature of about 48°F (9°C).

Test on Copper Coil Placed in a Water Tank

In this connection reference should be made to the test results shown in Table 4 where a coil consisting of 150 ft (49 m) of 1 in (2.5 cm) overall diameter was placed in a tank of water which was maintained at a constant temperature of 47°F (8.3°C) with antifreeze flowing through the coil at a rate of 2000 lb/h (909 l). The unusual value of heat transmission of 219 Btu/h/lin ft (65 W) was obtained. Variations in water temperature gave heat transmission values which were approximately proportional to temperature variations except that ice began to form on the whole length of tube surface when the water temperature in the tank fell below 39°F (4°C) and heat transmission fell to half the normal. The experiment was repeated so that the temperature of the water in the tank was not maintained. Heat transmission fell away as ice formed on the coils and was only 60 Btu/h/lin ft (18 W) with the water temperature at 36°F (1.7°C) after a 2 in (5 cm) diameter ice coating had formed on the tubes.

The potential advantage of having a large mass of water such as a swimming pool or large open tank becomes apparent. Heat is being collected in the open surface water mass from both solar sources and from ambient air and even a short length of coil can give very high rates of heat input. A pool 33 ft (10 m) in length by 13 ft (4 m) in width by 5.5 ft (1.5 m) in depth will hold 142,000 lb (64,600 l) of water. With a low-grade heat demand of 30,000 Btu/h (8800 W), the pool would meet this demand for 4.8 hours with a temperature drop of only 1 deg F. Alternatively, if the pool water was at 45°F (7.22°C), the demand could safely be met for 48 running hours which is approximately one week of average winter running.

A dwelling which had both heat pump and swimming pool would obviously arrange for both heating and cooling of the pool water from the heat pump, and the connections, etc. for extracting heat from or conveying heat to the water would be manual. Heating the water during the winter heating season may be necessary for both safety and storage purposes. Table 16 indicates that a heat pump to heat the building in Winter and the swimming pool water at other times would have a higher electrical annual load factor and a higher value of R_{η} than if the heat pump were used only for residential heating.

Solar Energy

The inclusion of solar energy in this chapter is a departure in that it is Available energy obtainable at a temperature higher than that of our ambient surroundings. But, like the pool of Unavailable energy around us, it is 'free' and there to be taken; like other sources of energy its capture involves a monetary cost in the provision of components. Yet, as has been shown, when used in a heat pump, the value of each unit of ambient energy may be at least equal to that of fuel.

At the best, solar energy is a variable quantity dependent on weather factors. Various estimates have been made as to solar energy intensity. I have been fortunate in being associated with Dr Farley in a long series of careful experiments made to measure solar intensity at his house in Norfolk. Careful recording was carried out during the latter part of the winter 1974 - 75.

From the 17th January to 31st March 1975, Dr Farley recorded solar heat coil outputs as high as 600 W/m^2 of coil area and over the whole period, which included a high amount of overcast skies, he estimated that the average collection over 24 hours was equal to 32 W/h/m^2 of coil area. Preliminary tests made by Dr Tovey at The University of East Anglia gave similar results.

This leads to an interesting possibility as regards the more efficient use of ground or auxiliary coils such as those referred to earlier and which are shown in Table 6. We know that a coil buried in soil gradually accumulated over the heating season a coating of ice which resulted in the antifreeze medium having a maximum incoming temperature of 32°F (0°C). The question that arises is whether a solar coil can feed sufficient heat, either into the antifreeze medium itself, or in some manner into the pipe system, at a temperature high enough to prevent the initial icing on the pipe surface in contact with the soil. If this were possible, a greater transmission of heat might be possible. The method using solar heat in the manner in-

46

dicated above could lead to an intriguing form of research and experiment. Alternatively the solar heat could be stored in an insulated tank or swimming pool.

Comparison of Ground Coil and Solar Heat Coil

A comparison between the two types of low-grade heat collection, in quantity and quality, will now be made. The buried coil in Table 6 has a length of 460 ft (140.3 m) and a surface area of 115 ft^2 (10.7 m^2). The coil collects solar heat at an average rate of 24,000 Btu/h (7034 W), i.e. 2240 Btu/m^2 (657 W) per hour, continuously over each 24 hours of the heating season but at a maximum temperature of 34.5°F (1.5°C), i.e. 53,760 Btu (15.77 kW) per 24 hours. We, therefore, have the comparisons in Table 6. If a buried coil is used, or a plastic coil on the surface of the earth, the source of immediate supply is really a mass of ice, which collects stored heat from the soil in the former case, and from ambient air in the latter case.

Table 6.   Table showing relative peak and mean daily heat in put to ground coils and solar coil.

Type of collector	Peak input per square metre of collector (W/h)	Average input per square metre for 24 consecutive hours (W/h)	Collection temperature (°C)
Direct solar	800–1000	32	4-25
Buried coil	657	657	0

Summary of Coil Components

Coil A is copper piping 470 ft (144 m) in length with an overall diameter of 1 in (2.5 cm). The coil has been in use for twenty years. The length is insufficient to meet a demand of more than 22,000 Btu/h (6450 W) at 34.5°F (1.5°C) and provides 46.8 Btu/lin ft. The cost in 1955 of materials and excavation was under £50. Experience shows that the spacing could have been reduced and the full heat requirements met with 750 ft (245 m) of coil laid in the same excavation.

Coil B is a small auxiliary coil of copper pipe 150 ft (45.7 m) in length inserted in a tank with dimensions 2 m x 1 m x 1 m. Some hot domestic waste water can feed the tank, and other feeds from gutters etc. are available; the coil normally provides 6000 Btu/h (1760 W). This coil can, if placed in a body of moving water, supply the whole demand of low-grade heat at a temperature of 40°F (4.4°C) if the water is maintained above 45°F, giving a value of 200 Btu/lin ft (58.6 W). Total costs of the installation were below £50.

Coil C is new and not yet fully tested. It consists of 360 yd (330 m) of black plastic hose pipe laid on soil and designed to take in solar heat by virtue of its black surface. The coil has not been made selective by glazing as it is not intended to run antifreeze flow at temperatures exceeding 40°F (4.4°C). Also it is designed to be free to any movement of ambient air. It is expected that an iced surface will be formed on the hose for much of the time that it is in use. The ice will absorb solar heat and heat from ambient air at temperatures exceeding 32°F (0°C) for probably 75 per cent of the winter heating season. Experiments made during the winter of 1974-75 would suggest that this coil may meet the required low-grade demands. The cost of supplying and installing the hose pipe and the crude wood container is £100. The operating winter temperature of the antifreeze liquid has yet to be ascertained and may be too low for efficient operation.

Coil D. In view of the inadequacy of Coil A to meet the demand of the larger compressor installed in 1961, a further coil of lead pipe, 500 ft (153 m) long, was installed.

Air as a Low-Grade Heat Source

The disadvantages of this source of heat were found (during my earlier experiments) to be:
(i) that as air temperature falls, the heating demand increases;
(ii) that for air with high humidity, icing of the evaporator occurred with ambient air temperatures as high as 48°F (8.9°C).

The use of air as the low-grade heat source will involve a larger evaporator than if liquid is used as the transfer medium from earth. The temperature drop in the air from which the low-grade heat is extracted must be kept reasonably low so as not to depress unduly the evaporator and suction temperature and pressure. Hence, relatively large volumes of air must be passed through or across the evaporator surface. Problems of noise may then arise and the fan power requirement also increases, thus reducing the overall P.E.R.

As the ambient air temperature falls, the demand for heat increases. Simultaneous with this demand for maximum output of heat at t_1, the value of t_2 is falling (see Fig.11) and the overall P.E.R. of the machine is being reduced. Any attempt to retain t_2 at a normal value would involve still larger quantities of air at a higher velocity, with the consequent increased fan power requirement.

There are several methods by which the two previously mentioned disadvantages of air as the low-grade heat source may be overcome. The findings of my experiments, which were discontinued through lack of funds, and augmented by the larger-scale work carried out during my collaboration with Messrs Lucas and Morris Motors led to the conclusion that the use of ambient air as the heat source, unless treated by one of the methods above, was likely to cause inefficiency of operation and a short life for the compressor (working on the vapour compression cycle). This view was supported in a recent paper by Goodall which refers to experimental work carried out by the Electricity Council at their Capenhurst Laboratory. The paper shows that, in the USA, compressor failures amounted to 28 per cent within five years of installation; fan motor failures and refrigerant leaks amounted to a further 15 per cent.

This record can be compared with my own compressor (using ground coils) which has run for 15 years without any failure or attention. Goodall also gives an efficiency curve for a heat pump using ambient air as the sole source of ambient heat; this curve is shown as a broken line in Fig. 14 and can be compared

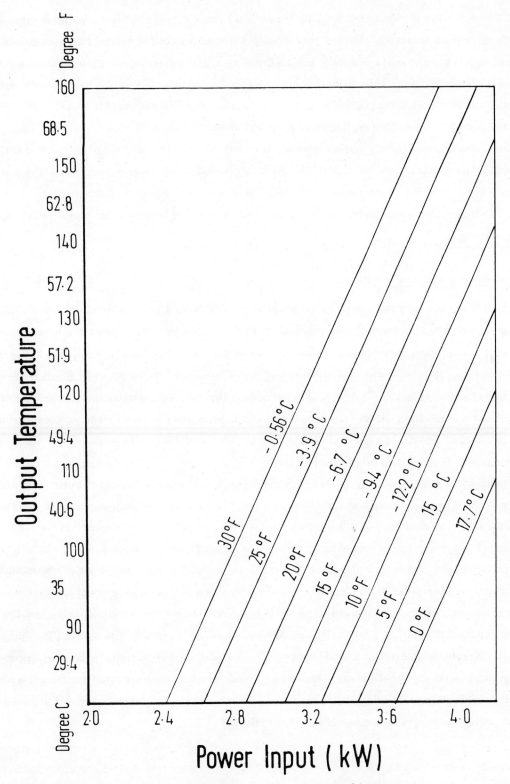

Fig. 11. Effect of compressor suction temperature on power input and heat output.

49

with the efficiency curve for my machine which uses a much more stable source of low-grade heat. It is exceedingly difficult to accurately size and load a compressor which must operate with a suction gas temperature that may vary between 14^oF and 50^oF (-10^oC and 10^oC). Undersizing causes intermittent overloading and short life; oversizing, while preferable, involves a lower value of R_η. The experimental work now being carried out by the Electricity Council is outdated by the very extensive experimental work on heat pumps carried out by Morris Motors, Messrs Lucas and myself since 1950.

The conclusion reached here is that ambient air is unsuitable for use with heat pumps when used, untreated, as the basic heat source. It is unstable and highly variable in temperature and humidity and requires unduly large volumes of air per unit of specific heat. On the other hand, ambient air can be of value when added to some other stable source of low-grade heat used as the basic source or if it is kept at a constant temperature.

Summary Regarding Low-Grade Heat

The source of low-grade heat for a heat pump should preferably be at a constant temperature throughout the heating season with a minimum temperature of 32^oF (0^oC). A large volume of water, or water drawn from the earth, will satisfy these requirements but may be expensive to provide. A coil laid in the soil will meet both requirements. Calculation and practice indicate a minimum heat output of 30 Btu/h/lin ft (8.92 W) at a temperature of at least 32^oF (0^oC). It is surprising that this cheap and simple method of obtaining low-grade heat has been ignored. An expenditure of about £300 to provide 1000 therms of constant temperature heat per heating season, at an annual cost of £0.015 per therm (one-tenth of the cost of heat from oil) would seem to be a reasonable commercial venture.

As the economics of heat pump usage convince more people, interest in the manufacture of small complete domestic heat pump units will be stimulated. Unfortunately, the tendency seems to be to follow the example of the USA in using air as the low-grade heat source, largely on the grounds that the air evaporator is cheaper to provide than a ground or air coil. As a result expensive supplementary electric resistance heating or reverse refrigerant flow devices have to be provided to meet those periods when ambient air temperatures fall below 40^oF (4.4^oC). The data already given in this chapter would lead to the view that the present unsatisfactory honeycomb-type of air evaporator is not necessarily cheaper than the much more stable method of installing a coil using soil and/or air as the low-grade heat source. Further experiments in this largely unexplored field would appear to be justified. There seem to be reasons to consider that design and experiment, when properly directed, should permit of an evaporator system using ambient air, which does not require de-icing, being produced. If such experiments were to succeed arguments as to the best source of low-grade heat would become irrelevant.

6: Refrigerants

Choice of Refrigerants

Before considering in detail the characteristics of the various refrigerants available, some knowledge of the upper and lower working temperatures, with their associated pressures, is necessary. One refrigerant may have a pressure widely different from another at a given temperature and the ratio of upper and lower pressures for a given temperature ratio may also be quite different.

Consideration of Most Suitable Working Temperatures

It has already been shown that the 'efficiency' of any heat pump is as follows:

$$R_\eta = \text{'Efficiency'} = \frac{\text{Heat delivered at } t_1}{\text{Heat equivalent of work put in}}$$

or

$$\text{Reciprocal Thermal Efficiency} = R_\eta = \frac{t_1}{t_1 - t_2} = \frac{q_1}{q_1 - q_2} \, .$$

This shows that in order to keep the value of R_η as high as possible, the difference between the upper and lower working temperatures $(t_1 - t_2)$ must be kept as low as possible. Since t_2 is largely fixed by natural conditions, t_1 becomes the variant term which must be kept as low as possible. It will also have been noted that $R_{\eta(I)}$ is a value for an ideal machine without any loss due to heat exchange, radiation or friction. The practical value of $R_{\eta(P)}$ will assume a 40 per cent loss so that $R_{\eta(P)} = 0.6 R_{\eta(I)}$, although good overall design may increase that value to $0.7 R_{\eta(I)}$ or higher.

To a great extent, the value of t_2, the lower working temperature, is determined for us so far as heat pump design is being concerned. It will be related to ambient air and water temperature which are the sources of low-grade heat. This can be related to the formation of ice coverings and will be considered here as being 32°F (0°C) and 273°K respectively, since we must use absolute temperatures. The upper working temperature (t_1) is also determined for us as will be seen from Table 7 which indicates the range of temperatures applicable to domestic heat pumps. Next comes the decision as to the temperature of the heat to be delivered to the rooms. This must generally be of the order of 100°F (38°C) if heated air is used

51

Table 7. *Table showing how the ideal and practical 'efficiency' of a heat pump falls as upper temperature increases lower temperature assumed constant at $T_2 = 20°F = 266°K$.*

T_1	(80°F) 299.7°K	(100°F) 316.8°K	(120°F) 321.9°K	(140°F) 333°K	(160°F) 344°K	(180°F) 355°K
$R_{\eta(I)} = \dfrac{T_1}{T_1 - T_2}$	9.0	7.0	5.8	5.0	4.4	4.0
$R_{\eta(p)} = 60\% \times \dfrac{T_1}{T_1 - T_2}$	5.4	4.2	3.5	3.0	2.04	2.4
Type of heating	Warm-air heating		Heated floors		Hot-water wall radiators	

and from $120°F$ to $180°F$ ($49 - 82°C$) if hot circulating water is used. A further factor in deciding the value of t_1 is the pressure/temperature ratio of the compressor and refrigerant used. It will be seen from Table 7 that the least efficient use of the heat pump will occur when wall or skirting radiators are used. The P.E.R. increases if floor heating is used; there is also a further advantage of the storing of heat available in the floor block with its large flywheel effect. The left-hand column of Table 7 shows the higher value of R_η when using the somewhat less efficient and probably less comfortable method of heating by warm air.

The pressure ratio for heat pump work will be related to the temperatures determined as shown above,

i.e. $\dfrac{180°F\ (82°C)}{20°F\ (-6°C)}$ to $\dfrac{100°F\ (38°C)}{20°F\ (-6°C)} = 9.7$ to 3.8 if CCl_2F_2 is used as refrigerant.

In refrigeration practice, steps would be taken, by means of water cooling, to restrict the upper temperature to, say, $96°F$ ($36°C$). For heat pump work refrigerants must meet higher temperatures without giving unduly high pressure.

Most Suitable Refrigerants

Liquids which are available as refrigerants for the vapour compression cycle are numerous, including water/steam. As many as thirty different liquids could be used. However, for reasons of toxicity and safe working pressures, etc. choice is limited to about five in the wide group of 'Freon' refrigerants. Insufficient attention has been given to this area of study. The rapid growth during recent years of the 'Freon' group of refrigerants suggest the possibilities of synthesising refrigerants to suit the particular purpose for which it is being used. A refrigerant suitable for refrigeration purposes is not necessarily the most suitable for heat pump purposes. The use of azeotropic mixtures introduces a possible field into physical chemistry for heat pumps which could produce safer and more efficient refrigerants based on their particular use; azeotropic developments of this nature have been under consideration in recent years.

Further study needs to be made of the different characteristics required of a refrigerant used in a heat pump from those required in a refrigerator. The operation DA (or da) in Fig. 7a is accepted in refrigeration practice as being by expansion through a throttling valve and the loss that the throttling operation causes is accepted as being inevitable. This leads to refrigeration practice considering that a refrigerant with a lower specific heat will be more likely to keep that loss to a minimum. But, heat pump practice favours the capture of this sensible heat (by subcooling) before returning it to the evaporator. Therefore, although a high value of the ratio, latent heat to sensible heat, is important, it could be less important as regards heat pump design.

Table 8 gives a range of four available refrigerants, showing the physical and thermodynamic properties of each. There appears to be no one selection that contains all the virtues what we would like. Very often it will be found that a desirable characteristic is cancelled out by an undesirable one.

The chief factors to be considered are:

(1) Evaporator and condenser pressures Table 8, column (a). Pressure in the evaporator should preferably be above atmospheric pressure and should not seriously exceed 200 lb/in^2 (gauge) in the condenser system. The lower the compressor ratio, the higher is the efficiency.

(2) Ratio of latent heat (L) to specific volume of vapour (V) Table 8, column (f). The ideal refrigerant will have a high latent heat, while the vapour will have a low specific volume at the evaporator and thereby decrease compressor work and size.

(3) Compression ratio Table 8, column (c). A low value is an advantage in that losses from clearance volume and leakage between cylinder wall and piston are kept low and the volumetric efficiency is likely to be better.

(4) Ideal horsepower required Table 8, column (i). With a theoretically perfect cycle all refrigerants would be equal in respect of this item.

(5) Ratio of latent to sensible heat (S) at t_2 and t_1 Table 8, columns (g) and (h). It is important to convert a maximum amount of liquid refrigerant into vapour at t_2 in the evaporator; a refrigerant with a high value of L/S if therefore preferable. The ratio L/S is considerably less (at t_1) in the condenser (col. (g)) and the sensible heat kept in the refrigerant after condensation may be 40 per cent, or more, of the total heat (q_1) that results after compression. Fortunately, as will be shown, much of this considerable quantity of heat can be collected and used within the cycle, either to raise t_2 or be added to the high-temperature delivery side of the system.

Vapour Volume at Suction Temperature Table 8, column (e)

The lower the vapour volume of the refrigerant as it leaves the evaporator and enters the compressor, the smaller the volumetric capacity of the compressor or, alternatively, the lower the speed at which the compressor can run.

Table 8 sets out the properties of the only four refrigerants in the Freon group suitable for our purpose. Each of the desirable characteristics shown above has been marked by a box. It is seen that Freon 12 (CF_2Cl_2) has the most 'merit' marks but is second to Freon 114 (CCl_2F_2) as regards latent heat at the evaporator. The disadvantages of Freon 11 become obvious. It has a volume at suction nearly eight times greater than Freon 12 and its suction pressure is sub-atmospheric. On the other hand, the use of Freon 12 at temp-

Table 8. Table showing various characteristics of four commonly used refrigerants.

	Pressure at −4°F		Pressure at 135°F		Pressure Ratio b/a	Latent Heat at 20°F (L)		Vapour Volume at 20°F (V)	Ratio L/V at 20°F	Ratio L/S		Ideal I.H.P
	p.s.i.a.	mbar × 10⁴	p.s.i.a.	mbar × 10⁴		Btu/lb	W/kg	ft³/lb		at 20°F	at 150°F	When O₂ is 24,000 Btu/h (7030 W)
Freon 12 (CF₂Cl₂)	21	1.47	205	14.4	9.95	80	23.6	1.121	71.5	5.4/1	1.02/1	2.0
Freon 11 (CFCl₃)	2.2	0.18	39	2.18	18	95	28.9	8.5	11.1	7.4/1	1.7/1	1.85
Freon 21 (CHFCl₂)	4.0	0.23	70	4.29	16.8	108	32	6.3	17	7.7/1	1.7/1	1.89
Freon 114 (CCl₂F₂)	5.0	0.3	80	5.5	16.0	61	18.0	2.1	30	4.4/1	1.19/1	2.03
	(a)		(b)		(c)	(d)		(e)	(f)	(g)	(h)	(i)

Table 9. Calculated throttling and excess work of compression losses for the principal refrigerants ($T_2 = 5°F$; $T_1 - T_2 = 80°F$).

Refrigerant	Work of compression for the Carnot cycle (Btu/lb)	Theoretical work of compression (Btu/lb)	Loss due to throttling process ab (Btu/lb)	Throttling loss in percentage of work in Carnot cycle	Excess work of compression	Excess work in percentage of work in Carnot cycle	Total losses in percentage of Carnot cycle
	(1)	(2)	(3)	(4)	(5)	(6)	(7)
Ammonia (NH_3)	84.06	99.6	7.7	9.16	7.8	9.28	18.44
Sulphur dioxide (SO_2)	24.97	29.07	1.84	7.369	2.26	9.05	16.42
Methyl chloride (CH_3Cl)	26.62	30.66	2.42	9.09	1.62	6.09	15.18
Dichlorodifluoromethane (CF_2Cl_2) Freon 12	9.19	10.85	1.61	17.52	0.05	0.544	18.06
Trichloromonofluoromethane ($CFCl_3$) Freon 11	12.0	13.40	1.30	10.83	0.10	0.83	11.66
Dichloromonofluoromethane ($CHFCl_2$) Freon 21	15.87	17.85	1.61	10.14	0.37	2.33	12.47
Dichlorotetrafluoroethane ($CClF_2$) Freon 114	10.7	11.93	1.23	11.5	0.18	1.7	13.2

eratures of 180°F (72.5°C) involves pressures of 300 p.s.i.a. which are relatively high for even a well-made refrigeration pipe junction. Reference is made later in the book to the use of 'cascading' in which a first compressor completes a cycle up to, say, 130°F (54.4°C) and a second compressor carries the cycle from this temperature up to, say, 190°F (87.7°C); in this case Freon 11 ($CFCl_3$) or Freon 114 (CCl_2F_2) could be used in the second compression system. Table 9 has been included to give a somewhat broader picture of the refrigerant range.

7:   Design project for a domestic heat pump

The Elements on which a Project should be built

With the information already given, a design project for a domestic heat pump can now proceed. The design data are related here to the particular case of a bungalow but it should be possible, in other cases, to modify each step or element to suit the special characteristics which the reader might encounter. The first three design steps will obviously be necessary in every case. In the last three steps entirely different decisions and conclusions might be reached from those given later.

A design project should proceed by considering each of the following elements:

Determination of the insulation value for the building.

Calculation of the heat loss for various climatic temperatures.

Decision as to the most suitable indoor temperature.

Decision as to the source, or sources, of low-grade heat.

Choice of the refrigerant to be used.

At this stage knowledge will have been gained as to the high-grade heat requirements for a given minimum ambient air temperature and a chosen suitable indoor temperature. We have, therefore, to find a value for q_1.

The use of either air or soil as the low-grade heat source should then be examined so as to determine the effect upon subsequent design of maximum/minimum operating temperatures and pressures. This will give two different cycles of operation and two machines, one using an air evaporator and the other a ground coil or other solar source. The subsequent steps are then:

Determining the heat values and vapour volumes at state points a, b, c and d in the cycle; subsequent plotting of state points on p/h chart (Fig. 35 in Appendix B).

Evaluation of ideal compressor, condenser and evaporator duties, etc.

Final evaluation of practical duties for components.

These last three steps involve calculations based on the use of p/h charts and standard refrigeration tables. Although the methods of using these charts and tables are explained, not all readers will wish to make use of them. For this reason, the above three steps have been abstracted from the main theme and placed in Appendix B. The remainder of the chapter, therefore, gives a summary only of the ideal design calculations for the two electric motor-driven machines using air and ground coil respectively; and also for the machine designed to the same output specification but driven by an oil engine.

STUDY OF BUILDING TO BE HEATED

Hot-Water Central Heating

The normal radiators used for central heating are designed for a given output of heat with water entering at 180°F (82.2°C). Output of heat falls off rapidly as the water temperature decreases. To heat water to the above temperature with a normal heat pump involves high pressures and low values of R_{η}; a two-stage heat pump (see Fig.18) would be advisable. Alternatively, floor heating gives a large equivalent radiator surface and a flow water temperature as low as 113°F (45°C) is permissible. This temperature is well within the range of a single-stage heat pump and permits a value of $R_{\eta} = 3$. It should be noted, however, that a properly designed radiator system can provide sufficient heating for 95 per cent of the heating season at temperatures well below 180°F.

Heating by Warmed Air

In this case the maximum air temperature required is between 125°F and 95°F (51.7°C and 35°C). Assuming a stable base temperature (t_2) by ground coil, etc., a high value of R_{η} can be achieved. This advantage may be lost if ambient air is used as the low-grade heat source unless one of several possible means are used to obtain stability of the low-grade heat source temperature.

Temperature Limits

It is preferable that the heating system should be designed for a maximum water flow temperature of the order of 140°F (60°C) when a heat pump is used. To obtain this maximum effect, therefore, the low-temperature radiant heating system, using heated floors, ceilings or walls, is the most suitable for use with a heat pump. The building chosen for this project, and for which the heat pump will be designed, is heated by means of warmed floors (low-temperature radiant heating system). Wall radiators can be used at this temperature of 140°F if the total surface area is increased proportionately.

Principles of Floor Heating System

With this system, the rooms are warmed by heating the floors of the structure instead of heating directly the air in the rooms. The warm floors emit radiant heat at a low temperature which travels upwards and radially in all directions. This radiant heat is partially absorbed and partially reflected at the surfaces affected. The reflected part is ultimately absorbed at some point of the room surface or by the contents of the room. All the reflected heat from the floors passes from surface to surface until this absorption takes place. Thus there is a process of radiation, then absorption, then of reradiation, until a tendency is reached towards thermal equilibrium, giving practically uniform conditions in all parts of the room.

Since complete thermal equilibrium is never reached, slight temperature differences occur between floors and ceilings, walls etc. These differences, though small, are sufficient to set up small but well-distributed convection currents which cause the air to be warmed uniformly, with a steady temperature balance.

In the bungalow which is the subject of study, all the rooms have fitted carpets, except for the kitchen and bathroom (tiled) and hall (carpet and parquet). Surface temperatures on the tiles of 85°F (29.5°C) and of 75°F (23.8°C) on the carpets provide a uniform air temperature of 68°F (20°C).

Application and Effect in Test Bungalow

Measurements have shown that, with double glazing and well-insulated ceilings) temperature differences throughout the carpeted rooms are less than 1 deg F and the differences between the carpeted and the tiled rooms are less than 2 deg F. As a result, draughts are unknown, no stuffiness noticed and high comfort conditions achieved. Ceiling heating was not considered because of the room height (8ft 9 in) (2.6 m) and the single-storey nature of the building. From experience, however, the comfortable feeling of one's feet and legs which results from warmed floors would, in my view, make floor heating (with carpeted floors) preferable to ceiling heating.

An alternative method of using heat pumps, for domestic heating, with an upper temperature limit of 120°F (48.9°C), is that of heating air which is supplied through ducts to the rooms in the building. Some economic advantage would occur, since the upper temperature limit, t_1, would be reduced from 140°F (60°C) to 120°F (49°C) as a maximum.

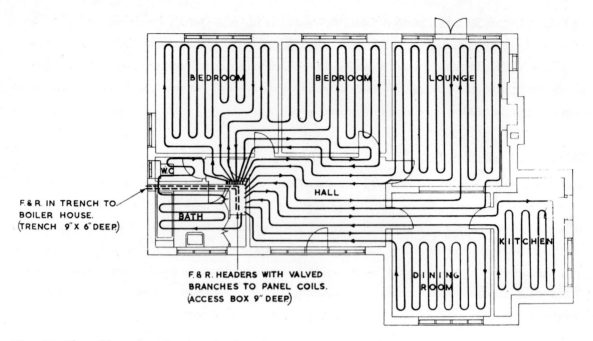

F & R. IN TRENCH TO
BOILER HOUSE.
(TRENCH 9" X 6" DEEP)

F. & R. HEADERS WITH VALVED
BRANCHES TO PANEL COILS.
(ACCESS BOX 9" DEEP)

Fig. 12. Plan of bungalow showing pipe layout.

Characteristics of Building to be Heated

The plan of the building is shown in Fig. 12 which also shows the pipe sections layout in the floor.

The concrete floor of the bungalow, which has an area of 1650 sq ft (153 sq m), consists of a slab 5 in (12 cm) thick, resting on a subraft of vermiculite concrete 6 in (15.4 cm) thick. The edge of the slab above the vermiculite is separated from the surrounding brickwork by 1.5 in (3.8 cm) of wood. Experience would suggest that the vermiculite slab can be dispensed with without serious heat loss downwards, provided that adequate edge-insulation is installed.

The bungalow has 11 in (28 cm) cavity brick walls (unventilated), a partially boarded loft and considerable window area. The windows and doors are extremely well-fitting and it is considered that, because of the absence of any draught-producing fuel appliance, the rate of air change is less than one house-volume per hour. The accommodation consists of a dining room and living room, two bedrooms, a large square hall similar in area to the rooms, a kitchen, bathroom and a workshop.

Distribution of heat is by means of three-eighths of an inch bore copper pipes buried at a depth of 2 in (5 cm) from the upper surface of the concrete floor slab. These pipes are spaced 12 in (30 cm) apart, and each room circuit is fed separately from a main header system fitted with sets of isolating valves.

Measurements of Building Heat Loss

In 1953 two improvements in heat insulation were made. Firstly, insulating slabs of expanded rubber 1 in

(2.5 cm) thick were laid over the ceiling joists. Secondly, internal windows were fitted providing double glazing with a $\frac{3}{4}$ in (1.9 cm) air space between panes of glass. These internal windows consist of glazed wooden frames screwed on to felt fixed on the wood rebates containing the outer casement windows. Table 10 shows the calculations for heat loss before and after these improvements in insulation.

Table 10.   Analysis of heat loss for bungalow subjected as study.

	Area, ft^2	U	Loss with original insulation, Btu/h°F	Loss with 1954 insulation, Btu/h°F
Window area	264	1.0 0.8*	264	160
Wall area	1 140	0.29	330	330
Floor and ceiling area	1 623*	(0.3 (0.16*	430†	225†
Volume ft^3	13 400	—	—	—
Convection loss	—	—	270	180
Floor loss	—	0.15	208	208‡

* Estimated revised value of U for better insulation at windows and roof.
† With 2 x ½ bedrooms not heated, the total heated area is reduced from 1625 ft^2 to 1450 ft^2 — calculated heat loss is then reduced from 1502 Btu/h°F to 1300 Btu/h°F. With improved insulation the net calculated heat loss is then reduced from 1103 to 950 Btu/h°F.
‡ Doors sealed with phosphor-bronze strips.

The detailed calculations as to heat output from a heated floor prove very complex. The heat-storage value of the extremely large mass of concrete in which the sinuous hot-water pipes are buried is high, and the 'condition' of the heat-storage block would appear to be some complex function of its temperature and of surrounding conditions. The most direct method of approximating the heat loss of the building and the heat output of the floor is the practical one of ascertaining whether the design condition of 68°F (20°C) in-door air temperature can be met with an outdoor air temperature of 28°F (-2.2°C). The calculated heat loss for the design conditions (see Table 10) would be 950 x (68 - 28)°F/h, i.e. 38,000 Btu/h (11,410 W), but a study of the ambient temperature conditions should be made to determine the lower temperatures.

Climatic Conditions

A knowledge of the climatic conditions is necessary in order to choose a minimum design temperature. If the machine is to be used for summer cooling, a study of wet and dry bulb and maximum summer tempera-tures would be essential.

For the type of heating postulated, however, the thermal 'flywheel' effect of the large heated slab of concrete is considerable and the design loading of the heat pump equipment will be affected very little by air temperature fluctuations of short duration. Minimum air temperature calculations can, therefore, be based upon the minimum temperature likely to be sustained for several successive days.

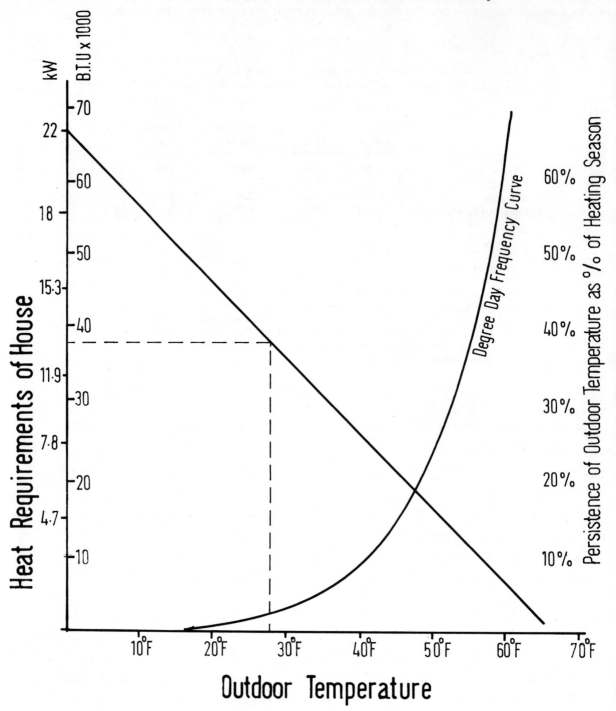

Fig. 13. Building heat loss and temperature frequency curves.

Reference should now be made to Fig. 13 which shows the heat requirements of the building related to outdoor temperatures. It also shows a part of a degree-day frequency curve for the heating season from mid-September to the succeeding mid-April. To meet a load covering 100 per cent of the heating season, the heat pump must obviously be designed for an output (q_1) of 68,000 Btu/h (19,930 W), but to meet the requirements covering 97.5 per cent of the heating season, the design load (q_1) can be reduced to 38,000 Btu/h (11,140 W). It is also known that the mass in the concrete floor will contain an amount of heat that will meet a temperature fall in the house of 0.21 deg F per hour and would therefore hold the indoor temperature so that it would only fall from $68°F$ $(20°C)$ to $63°F$ $(17.2°C)$ in 24 hours. This reasoning would suggest that the maximum heat input could be sized at about 38,000 Btu/h (11,140 W). If, at this stage, an assumption is made that a heat pump with an evaporator temperature of $28°F$ $(-2.22°C)$ will have a P.E.R. of $R_{\eta(P)}$ = 3, then it will be seen that the power required to drive the machine will be 38,000 ÷ 3 = 12,700 Btu/h (3722 W), i.e. 3.73 kW or 4.97 B.H.P. In fact, the present machine has a heat output of 42,000 Btu/h and a demand of 4.2 kW.

There had been some previous verifications of the heat requirements during the first winter of occupation (1950/51) before the first experimental heat pump had been completed. A 1000 gal (4543 l) tank, well insulated, was fitted with 13 kW immersion heater elements with the water heated to a maximum of $145°F$ $(62.8°C)$ controlled by the master thermostat in the house. The consumption during the year, with the thermostat set at $68°F$ $(20°C)$, was 35,800 kWh for a heating season of 3000 hours representing an average consumption of 11.93 kW/h = 39,360 Btu/h.

When considering the use of air as the low-grade heat source, allowance has to be made for the lower suction temperatures that will occur. Fig. 13 indicates that for probably 3 per cent of the season air temperatures below $30°F$ $(-1.11°C)$ will be encountered giving suction pressures (using CCl_2F as the refrigerant) as low as 15 p.s.i.a. (1.02 bar). This severe drop in suction pressure reduces considerably the output and P.E.R. of the machine (see Fig.14) and it was found that, under these extreme conditions, the value of $R_{\eta(P)}$ for the first experimental machine fell to 1.5. It will be uneconomical to size a simple air/evaporator machine to meet these low-temperature conditions. Design for both air and ground coil machines in this project will therefore be based on the design figure of 38,000 Btu/h (11,140 W) for q_1 during minimum outside air-temperature conditions.

Disadvantages of Air as Low-Grade Heat Source

The experience of using air as the low-grade heat source led to the installation of the ground coil shown in Figs. 9 and 10. Fig. 14 indicates how quickly the value of R_η falls away at evaporator temperatures below $32°F$ $(0°C)$, i.e. 44.7 p.s.i.a. A study of refrigerant tables for CF_2Cl_2 shows that between $25°F$ $(-4°C)$ and $0°F$ $(-17°C)$ the pressure ratio has increased seriously and the vapour volume has increased by a factor of 1.6. Even though icing troubles were less serious the other factors had seriously reduced the value of R_η when evaporator temperatures fell to $5°F$ $(-15°C)$. Such expedients as reversing the refrigerant flow

61

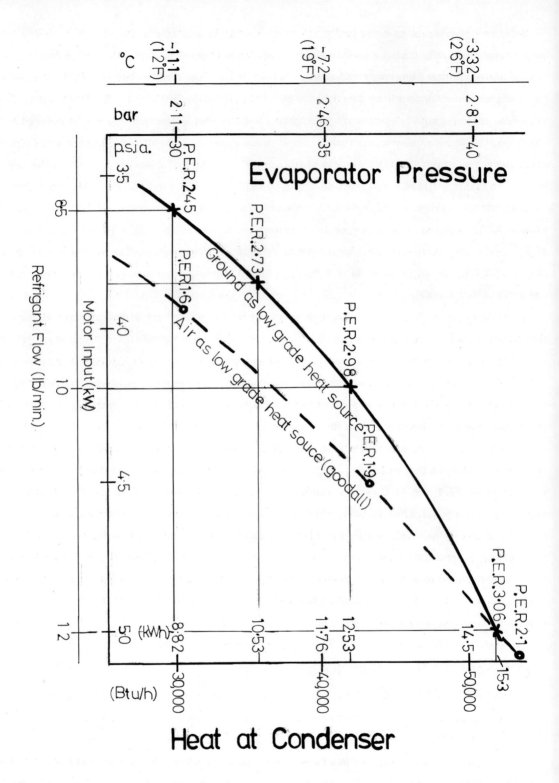

Fig. 14. Curves comparing and showing the variations of P.E.R. with suction temperature and pressure for the author's and Goodall's heat pumps.

or using the supplementary electrical resistance heating are a gross misuse of energy. Experimental time and money spent on testing these crude expedients should rather be spent on more promising methods, such as the use of the heat pump principle to increase incoming air temperatures and the use of the Ranque tube.

Since we are designing here a heat pump that must provide a specified amount of heat (38,000 Btu/h) under all atmospheric conditions, the evidence is that the simple air-evaporator heat pump cannot meet the specification. However, this is not to say that a satisfactory machine cannot be designed and, indeed, the evidence suggests that it can be designed.

Most Suitable Source of Low-Grade Heat

Since this subject is an essential factor in a design project, a brief analysis will now be made of the various alternatives discussed earlier.

Primarily, it seems to be a matter of cost and convenience of installation. We have to choose between either 'untreated' ambient air, which is unsatisfactory at times and leads to gross oversizing of plant or direct or stored solar heat.

Chapter 5 shows that 330 m of copper tube buried at a depth of 1 m in soil can use solar energy stored in the ground so as to provide sufficient heat (q_2) for our specification throughout the heating season at a steady temperature of 32°F (0°C). Evidence (yet to be proved) led to the tentative hypothesis that a length of 330 m of black plastic coil laid either on the ground surface, or just below the surface, even though it iced up for a part or even all of the season, might provide a steady seasonal source of low-grade heat, although probably at temperatures below 32°F. If this hypothesis proves to be true, then a satisfactory source of low-grade heat, largely independent of ambient air temperature, could be provided for £60 (at March 1975 prices) and an area of land equal to 150 sq m. Direct solar heat might augment this temperature during daylight hours. It has been estimated that half a million homes have been equipped, since 1933, with standby oil-engine driven generating sets varying from 2 to 6 B.H.P. Each of these engines, when coupled to a heat pump, can provide, from rejected fuel heat, at least 80 per cent of the low-grade heat requirements, per horsepower installed (see Appendix B, etc.).

Until, and unless, research and design can economically arrange for ambient air to be 'treated' so as to maintain a low-grade heat supply with a constant incoming temperature of 35°F (1°C) for the entire heating season, the future successful domestic heat pump is likely to favour stored and direct solar heat rather than ambient air as the low-grade heat source.

Most Suitable Room Temperatures

Comfort conditions for room heating are not easy to discover or to standardise. The conditions are dependent both on the likes and dislikes of the individual and also the active or passive state of the individual at a given time. The age of the occupants of the dwelling is also relevant in that a person aged 35 might well be comfortable at a thermostat setting of 65°F $(18.3^{\circ}$C) but most occupants having twice that age would not be comfortable below a setting of 70°F $(21^{\circ}$C). Annual consumption of 12,000 kWh represents a 70°F setting.

The cost of comfort, for the test building now being considered is fairly high, even though annual heating costs using a heat pump is only of the order of 35 per cent of the cost when using normal fuel. Tests made since 1961 would suggest that consumption increases by an average of 6 per cent for each 1 deg F increase in thermostat setting between 60°F and 68°F.

The following standards have been used for varying test periods in order to ascertain the psychological needs of the two occupants:

(a) Thermostat set at 68°F (differential = 0.5 deg F).

(b) As above, but thermostat set down to 58°F from 10 p.m. to 6 a.m.

(c) Thermostat set at 64°F.

The following observations apply:

(a) The highest standard of heating $(68^{\circ}$F – 67°F) is found to result in higher temperatures than some guests prefer. In view of the fact that the whole house is maintained at an average of 70°F day and night throughout the heating season, it represents a standard higher than that in use in the majority of houses in the country. To maintain the same standard by means of coal fires, in the test bungalow, would require four fires burning constantly with, of course, a very considerable variation in room temperatures.

(b) This does not permit the high standard of (a) to be achieved. In cases where the thermostat would normally be switching the heat pump on at about 10 p.m., the overall delay in resuming heating may be as much as 12 to 16 hours, which is greater than the 'comfort hold over' period of the floor. There appears to be no appreciable economy achieved by these lower night settings.

(c) This is the temperature $(64^{\circ}$F) originally adopted in 1961. Practically all the male visitors find the rooms quite comfortable for sedentary use without any auxiliary heating but some female visitors prefer a higher temperature.

The scope and nature of this subject is too wide to permit anything more than an account of the experience obtained during the tests and two points would summarise this experience. Firstly, it was generally found that, once an agreement as to 'warmth' was established, the minimum air temperature which established this feeling was found to be 'more comfortable' than a higher temperature. Secondly, agreement as to 'warmth' was found to be obtained only when circulation of warm water was taking place in the floor coils and the floor surface temperature was not of a lower order than 70°F. Given this condition, room air temperatures as low as 60°F $(15.5^{\circ}$C) were very often acceptable to occupants sitting still in the room. It was also found that the use of an imitation coal fire (electric), but without heating elements, or even the sudden incidence of sunlight into the room gave an acceptable feeling of 'warmth' at room air temperatures several degrees Fahrenheit lower than was possible without these psychological aids.

Horsepower of Heat Pump Related to House Size

We can, to some extent, anticipate the later calculations by summarising the data now available and relate heat pump requirements to house size. It must be emphasised that the calculations now to be made, and this also applies to the curve in Fig. 15, can only give approximate values. Each house has its own special characteristics and variations, e.g. room heights, exposure to wind and sun, insulation values etc. The final sizing of a heat pump can only be achieved from detailed calculations for each dwelling.

Within these limits we have found that a bungalow with an area of 1500 sq ft (140 sq m) and volume of 12,000 cu ft (340 cu m) would require a heat pump capable of a heat output q_1 of 38,000 Btu/h (11,140 W) for 70°F setting and 28,000 Btu/h (8220 W) for 65°F thermostat setting. A value of $R_\eta = 3$ for the heat pump has been adopted in these calculations. In my view it is possible to mass produce cheaply a heat pump having an overall seasonal value of $R_{\eta(P)} = 3$ for the conditions considered here and heat pumps with any lower value should not be used. By taking the above values it becomes possible to construct a curve (Fig. 15) which relates power requirement to house size and which gives an approximate guide to heat pump sizing.

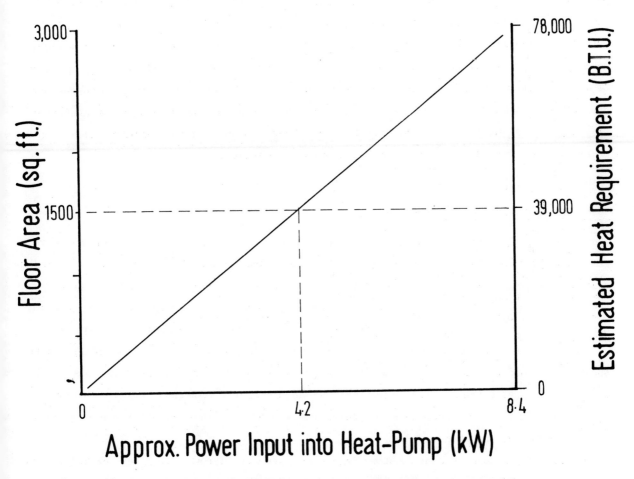

Fig. 15. Curve showing approximate relationship of heat pump power requirements and floor area of dwelling.

65

Sources of Low-Grade Heat

The choice of low-grade heat source lay between the use of ambient air or solar heat and the limitations of each system were stated in Chapter 4. For the purposes of the experiment ambient air was used during the winters of 1951-53 and a ground coil placed 3 ft deep in the ground in 1954 has been used since.

With the air/water machine, icing of the evaporator surface will occur under certain conditions and high-grade energy (taken either from the condenser or from an exterior source) must be degraded to avoid or periodically remove the icing. The earth/water machine needs no such provision.

The serious disadvantage of low suction temperatures, as illustrated in Figs. 11 and 14, is shown to be that, at any suction temperature below 15°F (-9.4°C), the required condenser temperature of 125°F (52°C) will not be reached. Also power input W_c falls off at the rate of approximately 0.1 kW per 1 deg F reduction of gas temperature in the evaporator, which represents a falling off in heat output q_1 of the order of 1000 Btu/h (300 W) per 1 deg F. Figs. 11 and 14 lead to the conclusion that the use of ambient air as the low-grade heat source involves the need (after allowing for a 10 deg F (5.5 deg C) drop in the evaporator between incoming air and refrigerant) to maintain the incoming air at a minimum temperature of 30°F (-1.1°C) if the required higher temperature and output are to be maintained. Nor does this allow for air humidity causing icing up of air passages. Therefore, either by recycling heat between condenser and evaporator, or by some form of wasteful supplementary heating, or by both methods, ambient air temperatures below 30°F (-1.1°C) and, indeed, probably below 40°F (4.4°C) will require an amount of higher-grade heat at the evaporator. In practice the increased suction gas volume at lower suction temperature causes a falling off in volumetric capacity and in the output per kilowatt of power input.

Use of Air or Radiators as the House Heating Medium

Not all readers will desire, or be able, to use a heated floor or wall surface. Some notes are therefore given as to alternative methods of distributing heat throughout the dwelling. Whether warmed air or radiators are used, good pump design postulates a hot-water calorifier placed in parallel with the heat pump condenser. A storage cylinder, i.e. calorifier, is as necessary to the efficient design and working of a heat pump as is the flywheel to an engine or a stable source of low-grade heat to the heat pump evaporator.

The calorifier should consist of a lagged hot-water tank of, say, 17.5 cu ft (0.5 cu m). Limit switches should be provided which can be set to switch the heat pump on and off at about 10 deg F (5.5 deg C) below or above some predetermined tank temperature. This tank can also be arranged to supply hot water for baths, washbasins and for general domestic use. Overall control of house temperatures can be by means of an overriding thermostat in the house.

If warm air is to be used as the house heating medium, the calorifier will have tubes inserted through which air is blown before entering the rooms. The design of a suitable warm-air system is beyond the scope of this book and reference should be made to standard textbooks.

If hot-water radiators are to be used with a heat pump using normal spacing and surface area, problems arise. A normal type of wall radiator has a heat emission of 500 W/sq m when supplied with hot water at 180°F (82°C). Therefore, to provide 38,000 Btu/h (11,140 W) 22 sq m of radiator is required. The condenser pressure of a heat pump (using CCl_2F_2 as refrigerant would then need to be over 300 p.s.i.a. (20.4 bar) which is too high. To supply hot water at the maximum heat pump temperature of 120°F (61.6°C) would reduce the emission from the radiators from 500 W/sq m to, probably, 330 W/sq m, i.e. giving a total heat emission with 22 sq m of radiator surface of only 7360 W instead of 11,140 W. Reference to Fig. 13 suggests that this lower emission would meet the house heat loss for ambient temperatures down to about 35°F (1°C) and would meet requirements for at least 90 per cent of the heating season. Alternatively, the radiator surfaces must be increased by about 30 per cent.

SPECIFICATION FOR FINAL DESIGN

The results from this study of design factors will now be put in the form of a final specification for two heat pumps, each driven by an electric motor.

Heat Pump A (Ideal Working)

A ground coil is to be used from which an anti-freeze solution enters the evaporator with a mass flow of 200 gal/h at 34°F (1.1°C) and leaves at 22°F (-5.6°C). It is assumed that the refrigerant is Freon 12 (CCl_2-F_2) and that the condenser and evaporator pressures are 195 p.s.i.a. (12.5 bar) and 41.6 p.s.i.a. (2.83 bar) respectively.

Heat Pump B (Ideal Working)

Air is assumed to be the low-grade heat source giving evaporator and condenser pressures of 31.8 p.s.i.a and 195 p.s.i.a. respectively, although in practice the condenser pressure would be lower than 195 p.s.i.a.

Final Ideal Heat Pump Design

The method used for design was briefly indicated in the second worked example shown earlier in Chapter 4. Not all readers of the book may desire to study closely this design method. The detailed design work for these two heat pumps is therefore carried out in Appendix B. For convenience, however, the heat quantities, etc. involved and the horsepower, etc. for each machine is shown in Table 11.

Table 11. *Table showing ideal cycle values derived from P/h chart and/or Refrigerant Tables for two heat pumps, A and B.*

Ground Coil	Air Evaporator
Anti-freeze enters at 34° F and leaves at 22°. Average suction temperature = 28° F. Suction pressure = 41.6 p.s.i.a. Condenser = 130° F = 195 p.s.i.a.	Air enters evaporator at 30° F giving suction pressure of 31.8 p.s.i.a. and 14° F Condenser = 130° F = 195 p.s.i.a.

Q_1 $= h_C - h_D$ 92 $- 38.7 = 53.3$ Btu/lb   92 $- 38.7 = 53.3$ Btu/lb

Q_2 $= h_B - h_A$ 81.4 $- 38.7 = 42.7$ Btu/lb   79.8 $- 38.7 = 41.1$ Btu/lb

W $= h_C - h_B$ 92 $- 81.4 = 10.6$ Btu/lb   92 $- 79.8 = 12.2$ Btu/lb

q_r = Refrigerant Mass Flow

$= \dfrac{38000 \times Q_2}{60 \times Q_1} = 11.88$ lbs/min $\dfrac{38000 \times 41.1}{60 \times 53.3} = 11.43$ lbs/min

I.H.P. $= \dfrac{W \times q_r}{42.42} = \dfrac{10.6 \times 11.88}{42.42}$ $\dfrac{12.2 \times 11.43}{42.42}$

$= 2.97$ I.H.P. $= 2227$W $= 3.49$ I.H.P. $= 2470$

P.E.R. $= \dfrac{Q_1}{W} = \dfrac{53.3}{10.6} = 5.02$ (Ideal)   P.E.R. $= \dfrac{53.3}{12.2} = 4.36$ (Ideal)

Notes on Table 11.

An approximate value of superheat has been added of 1.5° F. The heat values for the temperatures adopted, also pressure ratios etc are approximate only.

8:   Recycling waste heat

Compressor design for heat pumps differs considerably from compressor design for refrigeration. In general a heat pump compressor must arrange for the upper and lower temperatures to be as high as possible, whereas the refrigeration designer wishes to obtain the lowest possible values for t_1 and t_2. In both cases, of course, the design attempts to keep $t_1 - t_2$ to a minimum. Several differences in design may be quoted:

(a) In order to keep t_1 as low as possible, the refrigeration designer will not consider useful any heat which may arise during compression and by friction. On the other hand, the heat pump designer, not only uses the waste heat equivalent of the work of compression and friction but will also carefully collect any other heat losses in the circuit and return them to the system as low-grade heat which will lift the value of t_2 as much as possible. This course would be anathema to the refrigeration designer. An interesting example of such heat recovery in the heat pump circuit is the case where one of the earlier semihermetic compressors used by myself to operate my first experimental domestic heat pump was partially immersed in a tank. Arrangements were then made for the antifreeze solution, drawn from the ground coil which represented the low-grade heat source, to flow around the body of the compressor. The compressor body, which without such cooling settled down to a temperature of 168°F, increased the temperature of the antifreeze medium flowing at the rate of 2000 lb/h by approximately 1 deg F, representing an amount of useful recovered heat of the order of 2000 Btu/h (586 W). It was estimated that complete immersion in the tank would have increased this value to 3000 Btu/h (879 W). The use of completely hermetic compressors for heat pumps which, including the electricity supply terminals, can be completely immersed in water should facilitate to an even greater degree such conservation and use of heat losses in the heating machine.

(b) A further example of the difference in design outlook occurs when considering the expansion operation DA in Fig. 7a. For the conditions considered here the enthalpy of the liquid refrigerant after condensation will be of the order of 39 Btu/lb and this value must fall to approximately 12 Btu/lb before entering the evaporator. Since we are considering a refrigerant flow of 12 lb/min, the difference in heat values will be about 27 x 12 lb x 60 min, say, 19,000 Btu/h (5697 W). In the heat pump cycle this relatively large quantity of heat can normally be put back, by subcooling, into the low-grade heat circuit so as to lift the value of t_2 and thereby increase the general efficiency of the cycle. But in the pure refrigeration process, the prime function of which is to keep t_2 as low as possible, this method of disposing of the specific heat would not be justified even though it might increase the refrigeration effect.

Theoretical Aspects of Recycling Heat

Now let us consider how we can use this recycled heat and the effect that it will have upon efficiency. It was shown earlier that the 'efficiency' of the heat pump operation, as measured by the Performance Efficiency Ratio, R_η, is dependent upon the difference between the upper and lower working temperatures, i.e $t_1 - t_2$. If we can do anything to increase t_2 while keeping t_1 constant, we can reduce the ratio of compression (r) and therefore the work (w); and we have for an ideal cycle

(a) $\quad R_{\eta\,(a)} = \dfrac{t_1}{t_1 - t_2} = \dfrac{333^oK}{(333^oK - 273^oK)} = 5.55$ (when $t_1 = 60^oC$ and $t_2 = 0^oC$).

(b) $\quad R_{\eta\,(b)} = \dfrac{t_1}{t_1 - (t_2 + -t_2)} = \dfrac{333^oK}{(333^oK - 279^oK)} = 6.17$ (when $t_1 = 60^oC$ and $t_2 = 6^oC$).

(c) $\quad R_{\eta\,(c)} = \dfrac{t_1}{t_1 - (t_2 + - 2t_2)} = \dfrac{333^oK}{(333^oK - 285^oK)} = 6.93$ (when $t_1 = 60^oC$ and $t_2 = 12^oC$).

Those readers who have studied Appendix A or B should now turn to the exaggerated pressure/enthalpy diagram in Fig. 16, illustrating one effect of increasing the suction pressure and temperature $(P_4$ and $t_2)$. In cycle ABCD of Fig. 16 the value of R_η is

$$\frac{H_C - H_D}{H_C - H_B} = \frac{q_1}{\text{Work done at } P_4 \text{ and } t_2}$$

If, instead of using an expansion valve (operation DA) we could use an expansion engine, or by subcooling, we might capture the heat energy represented by area DAF. Assuming that this heat energy could be moved to area AA_1B_1B the suction gas pressure and temperature would be increased from P_4 and t_2 to P_3 and t_2'. The work area is thereby reduced from CBH to CB_1H_1 and

$$R_\eta = \frac{H_C - H_D}{H_C - H_{B_1}}.$$

Taking heat values for each of three cycles, we have:

For P_4 and t_2 = R_η = $\dfrac{72 - 36}{72 - 66} = 6$ $\qquad$ pressure ratio r = $\dfrac{P_1}{P_4}$.

For P_3 and t_2' = R_η = $\dfrac{72 - 36}{72 - 67.5} = 8$ $\qquad$ pressure ratio r = $\dfrac{P_1}{P_3}$.

For P_2 and t_2'' = R_η = $\dfrac{72 - 36}{72 - 69} = 12$ $\qquad$ pressure ratio r = $\dfrac{P_1}{P_2}$.

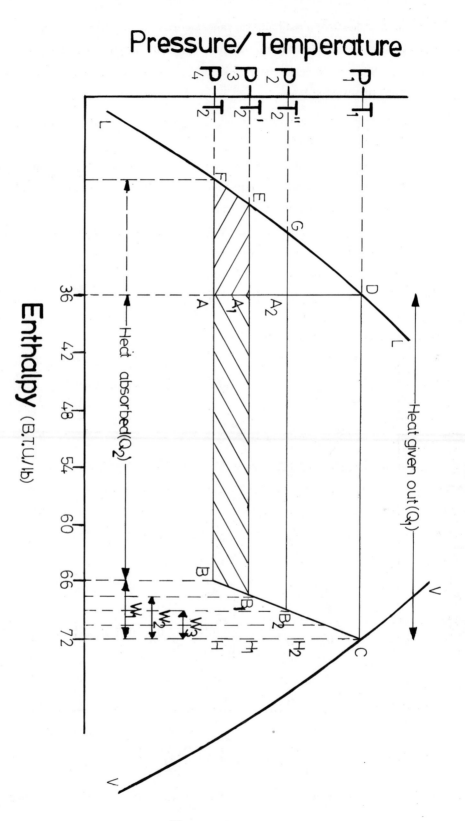

Fig. 16. Approximate p/h diagram to show how P.E.R. (R_η) varies with increase in suction temperature and pressure.

The recycling of heat referred to above causes certain other changes in the value of q_2 which are not shown directly in the specially constructed and exaggerated p/h and T/S diagrams of Figs. 16 and 17.

Methods of Recycling Wasted Heat

First consider the amount of waste heat which is, theoretically, available for recycling in the type of domestic heat pump being studied.

(a) The heat equivalent of the difference between the actual power used in compression and the ideal adiabatic compression (this will include electric-motor losses if such a motor is used for driving the compressor) is available and the ideal value of this heat will be, say,

$$(5.19 - 2.97) \text{ B.H.P.} = 2.22 \times 2545 \text{ or } 5650 \text{ Btu (1656 W)}. \quad \text{(see Table 11)}$$

(b) Next there is the sensible heat in the condensed refrigerant in the condenser between condenser and evaporator temperatures, less any undercooling which may take place in the condenser.

Let us look at the heat potentials when we have raised 12 lb x 60 = 720 lb/h (327 kg/h) of Freon 12

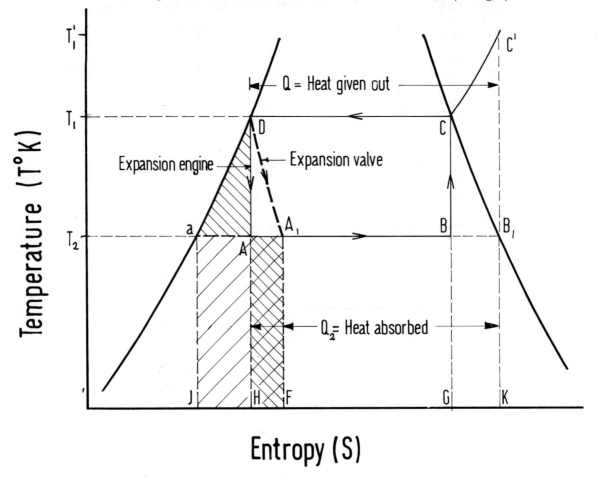

Fig. 17. Approximate T/S diagram to show effect of utilising condensate heat (subcooling) by recycling.

(CF$_2$Cl$_2$) from 20°F (-6.6°C) to 120°F (48.9°C). In condensing, the vapour will have given up 720 x (90.15 - 36.16) = 38,800 Btu (11, 375W) and the heat in the condensate = 720 x (36.16 - 12.55) = 16,990 Btu (1012 W). The cycle values are shown in the T/S diagram of Fig. 17. It will be seen from this figure, which is indicative only and does not truly represent the cycle, that, because we use an expansion valve, the condensate (with H = 36.1 at D) enters the evaporator at point A$_1$ with the same value of total heat of 36.16 Btu/lb (10.6 W). If we cool the liquid along line LL from D to a, the liquid would enter with a total heat of, say, only 12.55 Btu/lb (3.68 W) and it should be possible to utilise for other purposes within the cycle at least some part of this fairly considerable amount of specific heat which is approximately represented by area AA$_1$FH. It will be noted that the amount of heat q$_2$ taken in is increased by the ratio $\frac{BA}{BA_1}$. Appendix B contains further notes on subcooling.

Can this difference in temperature, pressure and heat value be usefully employed, e.g if the pressure difference was used to drive an engine or turbine? The mechanical problems involved are considerable because of sealing problems at the driving shaft, etc., and also the change of heat energy into kinetic energy would be at rather low efficiency because of the ratio of blade speed to refrigerant velocity. But, an effective value of expansion work of even 1 B.H.P. would increase the value of R$_\eta$ in Table 11 from 2.49 to 2.95, while an effective value of w$_e$ = 2 B.H.P. would increase R$_\eta$ to 3.58.

A further way to obtain recycling is to use the reversed method of 'feed heating' used with steam turbines in an attempt to approach the ideal Rankine cycle. There are several practical ways in which a reduction of the wasteful irreversible throttling via an expansion valve and a lower mean temperature of energy rejection of the condensate at t$_1$ may be achieved.

The compound arrangement (Fig. 18) could be used in cases where a higher value of t$_1$ was required,

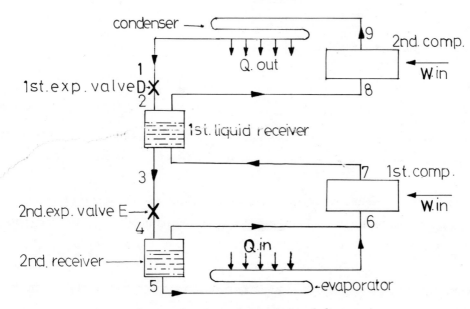

Fig. 18. Diagram showing two heat pumps in series (compounded).

73

e.g. to supply heat to a radiator system at $180^{O}F$ ($82.2^{O}C$). The first stage could be used for supplying domestic water requirements. There are two compressors, two expansion valves but, normally, a single condenser and evaporator. Overall P.E.R. is about 10 – 15 per cent better than for a single stage. Other similar alternatives are possible.

In the arrangement in Fig. 19 the liquid refrigerant from the condenser is passed through valve D into

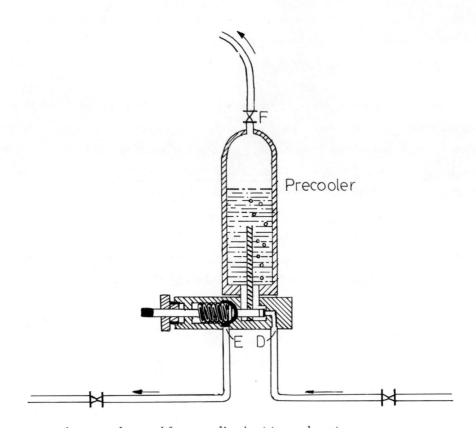

Fig. 19. Diagram of a precooler used for recycling heat to condensate.

the precooler at a lower pressure. A portion of the liquid evaporates, thus cooling the remainder, which collects at the bottom of the receiver and passes through valve E, either for further subcooling purposes or to the evaporator.

A further practical way of recycling is shown in Fig. 20 in which the heat exchanger GSX introduces superheat into the suction gas. Since this may introduce an undue amount of superheat into the compressor, it is possible to limit the amount of heat leaving this heat exchanger and to use a second one AFX which arranges for the balance of condensate heat leaving exchanger GSX to be used to raise the temperature of the antifreeze medium entering the evaporator. This is a reversed form of 'feed heating'. A more detailed study of the effects of subcooling is given in Appendix B.

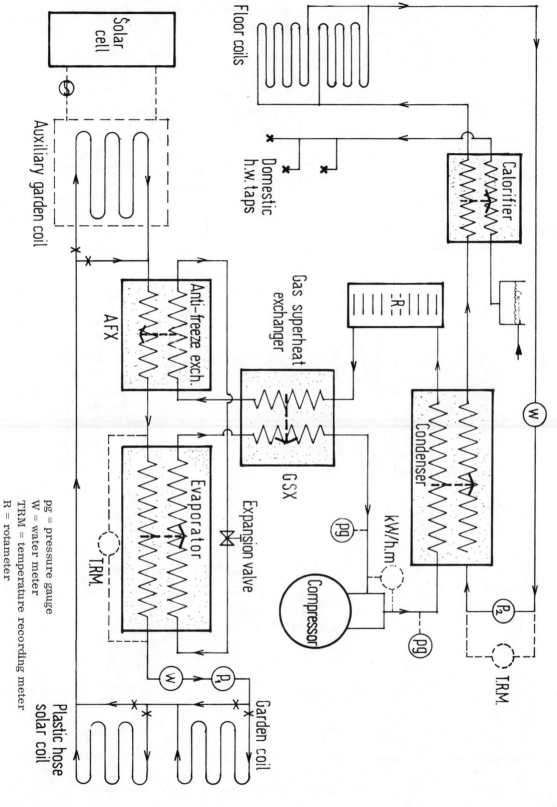

Fig. 20. Heat-exchanger layout for utilising heat in condensate.

pg = pressure gauge
W = water meter
TRM = temperature recording meter
R = rotameter

75

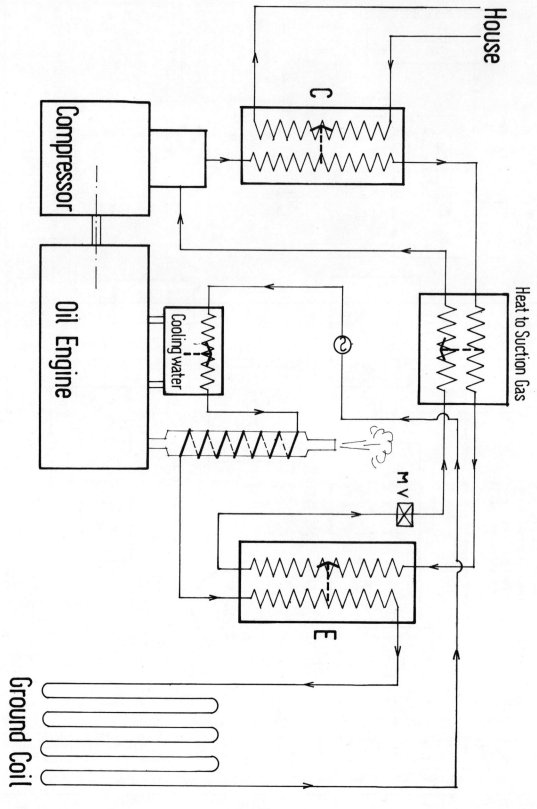

Fig. 21. Layout of an oil-engine driven heat pump assembly.

76

Oil-Engine Driven Heat Pump

A most interesting case of recycled heat is that of the oil-engine driven heat pump shown in Fig. 21. For this type of heat pump, the heat balance, with the value of components of heat available for recycling, are as follows:

(a) Heat initially in the fuel = 35,000 Btu (10,300 W).

(b) Heat available from the fuel as power = 4.86 B.H.P. (3626 W) = 12,372 Btu/h.

(c) Total heat output q_1 = R x 12,372 Btu/h = 49,500 Btu/h when R = 4.

(d) Heat loss in the engine = 65% x item (a) = 22,750 Btu/h.

(e) Heat available in condensate at condenser unit = 8.72 lb x 60 min x (41 - 19.17) Btu/h = 11,420 Btu/h.

(f) Total heat available for recycling = 34,170 Btu/h or 10,015 W.

Of the 22,750 Btu/h (7762 W) available from the engine losses, approximately 9240 Btu (2707 W) will be available from the engine cooling water which may be at a temperature of, say, 140°F (60°C). Depending on the upper temperature at which the heat pump is delivering heat (q_1), this recycled heat could either be introduced into the main output stream at t_1, or the heat could be made available as part of the low-grade heat input (q_2) by one of the methods shown in Fig. 21.

Next to be considered is the balance of the engine heat, i.e 13,510 Btu/h (3960 W) rejected in the exhaust gases. This heat will leave the cylinder at, say, 310°F (155°C) and could safely be cooled down to a temperature avoiding corrosion of the heat exchanger or undue back pressure if, say, 65 per cent of the heat was extracted so as to provide 8750 Btu/h (2574 W) at 200°F (93°C). Here again, there would be a choice between adding heat to the main stream (q_1) or using the heat quantity as low-grade heat at relatively high temperatures so as to increase the compressor suction pressure and thereby increase the cycle efficiency. The various alternative recycling routes are illustrated in Fig. 21. It is shown elsewhere that the oil-engine driven heat pump reduces the demand considerably for low-grade heat from other sources and increases the values of t_2 and R_η.

9: Modern heat pump practice

The first part of this chapter is largely historical. Details and illustrations (Figs. 22 to 25) are given of the second machine constructed by me and which was used from 1952 to 1961. I have not included design data as the need to design and construct individually is not so great now as it was in 1950. At that time, there was very little available in the way of compressors and components between 1h.p. and 10h.p. As Table 12 shows, that slot has now been filled and both compressors and components can be obtained in four or five steps within the above range.

A third machine (Figs. 26 and 27) was installed in 1962 replacing my second one, which was loaned for heating a swimming pool and is still in use. All parts of the new machine are of standard manufacture and it has been in continuous operation since its installation. It may be noted that no repair or opening up of any part of the plant has been necessary during the 14 years x 3000 h, i.e 42,000 hours of running, nor has any new or additional lubrication oil been used. This should meet the frequent criticism that maintenance costs and repairs on this type of rotating machinery are high. An annual check and topping up of refrigerant is all that has been necessary. An illustration of some of the measuring instruments is shown in Fig. 23.

Notes on Early Experimental Heat Pumps

The two machines, as illustrated, were made up in the form of prototypes. Figure 22 shows the air/water machine assembly excluding the air-type evaporator. The condenser consists of 6 in (5.1 cm) bore steel tube containing a motorcar radiator type matrix. Flanges were welded on to each end of the tube and ends bolted on as shown. The component to the front right of the picture is the superheater unit.

Figure 23 shows the air evaporator mounted on the wall, with connections through the wall to the machine parts mounted inside the building. Figure 24 shows the fan mounting and the machine when fixed inside the building.

At one stage of the experiments the complete heat pump was assembled and operated outside the building as shown in Fig. 25. The rectangular metal structure behind the heat pump was made up to prevent recirculation of the incoming air, the fan being mounted inside and drawing air in at a height of 8 ft, the air being discharged through the vent at ground level. The third, and present, machine is shown in Figs. 26 and 27. For purposes of historical record a photograph of the Norwich Heat Pump referred to earlier in the book is included (Fig. 28).

Table 12.   Characteristics of compressors.

Compressor no. of cylinders	2	2	3	3	3
Compressor bore	2 in	$2\frac{3}{8}$ in	2 in	$2\frac{3}{16}$ in	$2\frac{3}{8}$ in
Compressor stroke	$1\frac{7}{16}$ in	$1\frac{7}{16}$ in	2 in	2 in	2 in
Compressor swept volume	9.0 cu. per rev	12.7 cu. per rev	18.8 cu. per rev	22.6 cu. per rev	26.5 cu. per rev
Compressor pumping capacity	453 ft³/h	638 ft³/h	945 ft³/h	1132 ft³/h	1330 ft³/h
Compressor oil change	90 oz	90 oz	120 oz	120 oz	120 oz
Dipstick depth	Sight glass	Sight glass	Sight glass	Sight glass	Sight glass
Motor type single-phase	Capacitor start / Capacitor run	Capacitor start / Capacitor run	— / —	— / —	— / —
three-phase	Squirrel cage	Squirrel cage	Squirrel cage	Squirrel cage	Squirrel cage
Motor horsepower	2	3	4	5	6
Motor speed	1450 rev/min	1450 rev/min	1450 rev/min	1450 rev/min	1340 rev/min
Motor winding res. at 67° F run winding	0.9 Ω	0.6 Ω	—	—	—
start winding	5.0 Ω	4.0 Ω	—	—	—
Three-hase terminal to terminal	7.2 Ω	3.2 Ω	3.3 Ω	2.8 Ω	2.2 Ω
Motor overload code single-phase	MRA OE09	MRA OE09	—	—	—
three-phase	MRA OE09	MRAOE09	MRA OE09	MRA OE09	MRA OE09
Pipe size: suction	$\frac{3}{4}$ in o.d.	1 in N.B.	$1\frac{1}{4}$ in N.B.	$1\frac{1}{4}$ in N.B.	$1\frac{1}{4}$ N.B.
Pipe size: discharge	$\frac{3}{4}$ in o.d.	$\frac{3}{4}$ in o.d.	$\frac{3}{4}$ in o.d.	$\frac{3}{4}$ in o.d.	$\frac{3}{4}$ in o.d.

Fig. 22. Author's 1951 heat pump.

Fig. 23. Air evaporator as used, 1950-2.

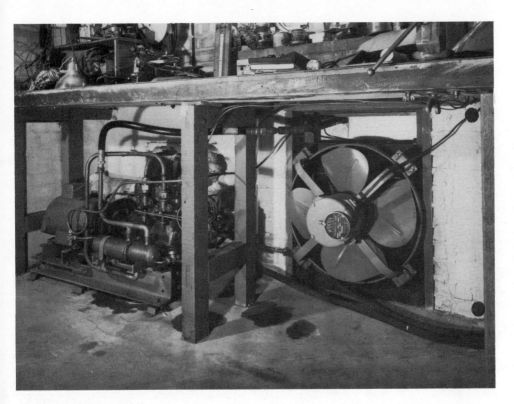

Fig. 24. Air evaporator assembly.

Fig. 25. Outdoor assembly using air evaporator.

Fig.26. Author's third heat pump,
using solar heat from coils.

Fig.27. The same heat pump,
showing instrumentation.

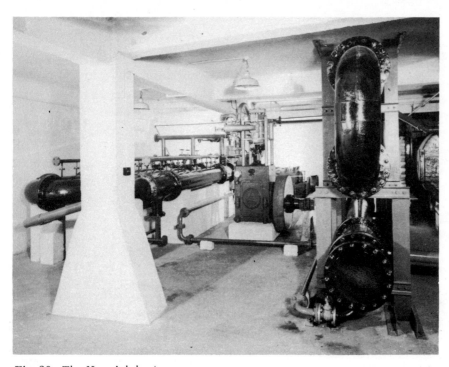

Fig.28. The Norwich heat pump.

Modern Practice in Heat Pump Development

Despite the title of this chapter there has been no growth, in Britain at least, in modern heat pump practice. In the USA the use of heat pumps has increased considerably and, consequently, there has also been some increase in the knowledge of the art. There are, indeed, some signs that progress in the USA has outstripped performance to the detriment of careful scientific development.

There remains a great opportunity in Britain, where the principle of the heat pump was first formulated, to make up for wasted years. At the risk of reiteration the two facts which constitute the basis and reason for this book are restated.

(1) Britain (and the world) is within measurable distance of the end of fossil-fuel supply.

(2) No known device, except the heat pump, can give a quantity of heat that is equal to, or may exceed, the heat value of the fuel consumed.

The rider may be added that the value of the heat pump must increase in inverse proportion to the quantity of remaining fossil-fuel.

Development of Technique

Technical development of the components which comprise a heat pump have been considerable during the past decade and have been due to progress in the refrigeration industry. It is remarkable that the sales and development effort associated with technical development appears to have been concentrated upon the use of improved and more efficient plant purely for refrigeration purposes.

Improvement in Compressors

The chief development here has been the change from the somewhat clumsy and inefficient separately driven compressor of Fig. 22 to the combined semihermetic compressor and electric motor with sectional details as shown in Figs. 29 and 30. It will be seen that the suction gas passes through the electric motor and windings and, thereby, takes up some part of the I^2R losses which occur in the electric motor. There is still no provision for recycling some part of the heat loss between polytropic and adiabatic compression, and in friction. As shown by a test earlier in the book, this can amount to at least 586 W/h. As suggested earlier, provision could be made in design which permits the whole motor-cum-compressor unit to be surrounded by the antifreeze medium which carries the supply of low-grade heat or by water-cooled cylinders. To a refrigeration compressor designer the heat quantities mentioned above have less significance than they would have to the heat pump designer. In the former case the lowest possible value of t_2 is being sought, and in the latter case the highest possible value of t_2 is the means by which the value of the Reciprocal Thermal Efficiency (R_η) is improved.

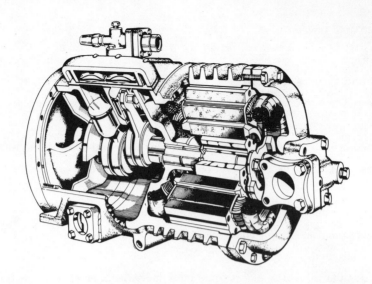

Fig. 29. Sectional view of modern semi-hermetic compressor.

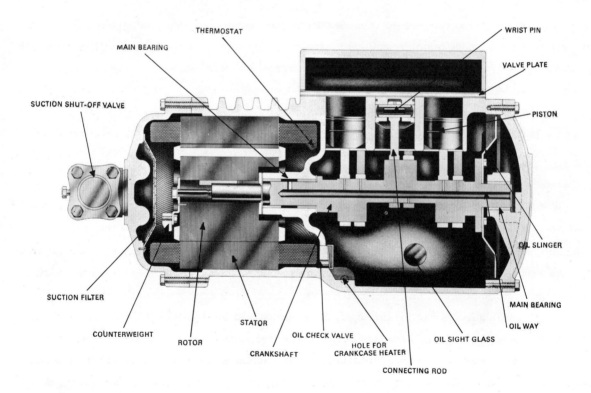

Fig. 30. Sectional view of semi-hermetic compressor.

84

The advantages and disadvantages of air as the low-grade heat source have already been discussed. The use of air for space heating purposes does have the advantage that, with the heat pump, the same equipment can be used for heating in the winter and cooling in the summer by reversing the refrigerant flow. The arrangement is shown in Fig. 31 which shows separate liquid receivers and piping for bypassing each exchanger.

This aspect of domestic heat pumps has undergone considerable development in the last two decades. In my view, this development has proceeded at the expense of research into the undoubted potentialities that this type of heat pump possesses and is still in the precondenser stage of the steam engine. The vapour compression cycle permits a value of R_η for the practical machine to have a value exceeding 50 per cent of the Ideal Carnot value, yet with the air cycle the practical value of R_η rarely approaches more than 20 per cent of the Carnot value.

Experiments and tests would suggest that the somewhat crude outline system shown in Fig. 31 could be considerably improved so as to give a winter seasonal value of $R_\eta = 3$ without recourse to supplementary electrical aid.

During the past twenty years no development in design of domestic heat pumps has taken place and the importing of electric motor-driven air-to-air machines manufactured in the USA has recently started. These machines are somewhat crude and unsophisticated. During adverse winter conditions their P.E.R. is understood to fall as low as 2 and a fairly large supplementary electrical load is embodied in their design. Messrs Lucas were making air-to-air heat pumps more than twenty years ago which were at least as good as these. A relatively small sum spent on research could result in a British product with a large world market. The manufacture of the ground-to-water heat pump with a constant seasonal P.E.R. of 3 could recommence and an air-to-air heat pump without supplementary electric heating and with a winter P.E.R. of 3, being a viable product, could be made available for manufacture. An oil-engine driven heat pump with a P.E.R. of 4 is an immediate practicality rather than a visionary proposition. The experience and scientific knowledge are available, and only awaiting direction and encouragement.

Improvements in Condensers and Evaporators

All improvements in the design and efficiency of condensers and evaporators are synonymous with improvements in both refrigeration and heat pump practice. In this respect such improvements do not match those made in compressor design during the last two decades.

Sectional details of a modern heat-exchange unit are shown in Fig. 32. Heat-exchangers, which at first seem to be the simplest component to design, are far from reaching the ideal for which one would hope. On the basis of heat transferred to energy dissipated in flow through the component there has been little improvement for many years. The Rosenblad type of heat-exchanger which gave spiral flow gave excellent results but has not been generally adopted. The unduly high temperature drop in most modern heat-exchangers, particularly in the evaporator, can reduce the value of R_η considerably, as was shown earlier in the book.

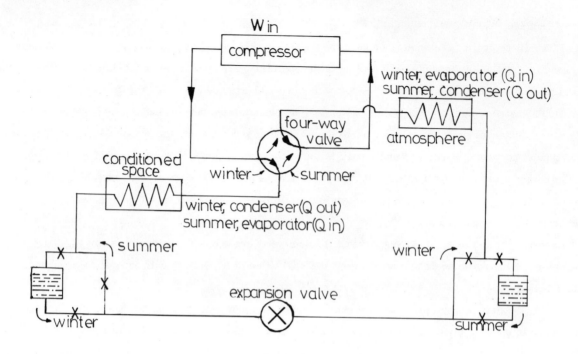

Fig. 31. Diagram of heat pump, taking heat from air with provision for reversing flow direction of refrigerant.

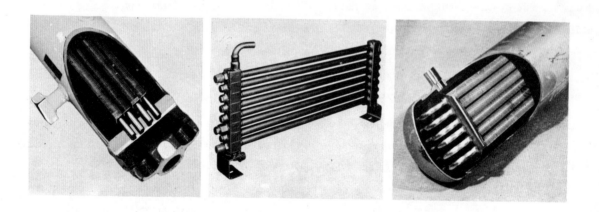

Fig. 32 Sectional view of heat exchanger.

How to Obtain and Specify a Heat Pump

Many correspondents have written to ask from which sources a ready-made heat pump can be obtained, suitable for use in domestic dwellings. The first step, as already pointed out, is to ascertain the heat loss of the house, either by making their own calculations, or by obtaining expert advice. The next, much more difficult, step is to find a manufacturer who can supply a heat pump sized to meet economically the specified heat losses. An alternative is for the enquirer to obtain the necessary standard components (based on design data given in this book) and to arrange to have these components assembled by skilled labour. Cost of machines purchased as a unit will be found to be high and probably of the order of £150 to £200 per brake horsepower. Quotations will usually be given for a standard refrigeration machine, without any subcooling or recycling equipment, from one of the two or three large firms making refrigeration plant. Imported machines are now becoming available using ambient air as the low-grade heat source and heated air as the house-heating medium and the problems related to these machines have already been dealt with.

It is suggested here that any specification for a domestic heat pump should incorporate a clause requiring a minimum seasonal value of $R_\eta = 3$ for a suitably sized electrically driven heat pump and a minimum seasonal value of $R_\eta = 3.5$ for an oil-engine driven heat pump. Until design and manufacture of heat pumps is dealt with nationally, on a mass production basis, it is unlikely that a manufacturer will be found who will comply with such a specification.

In 1975 I prepared a specification and marketing appreciation covering the field of heat pumps from 2 to 10 B.H.P. The response from two large British engineering groups who were approached was to the effect that it would involve alterations to plant and machinery that would not be justified unless an established market was available. After years of experiment and study, it is my view that hard facts will lead to the 'rejected stone', which is the subject of this book, becoming the 'head of the corner'.

Summary

This chapter may be summarised by saying that, firstly, modern heat pump practice is lacking in Britain and there is no modern specification available. All testing and research projects are scattered, independent and carried out on a part-time basis. Secondly, there are no facilities available to meet a sudden and growing demand for heat pumps, with the result that machines of unsatisfactory design are now being imported in increasing quantities.

In order to meet these deficiencies, I would suggest that the following steps be taken:
(1) The country has a declining motorcar industry, in which demand may decline at an annual rate of about 5 per cent. This industry has the necessary mass production plant and the labour and skills to produce a domestic heat pump at prices competitive with other forms of fuel-burning apparatus of equal capacity. An allocation of 1 per cent of £1600 million currently being invested in the industry should be used for research into, and manufacture of, a mass-produced heat pump for domestic use with a minimum P.E.R. of 3.

(2) Research into the design and manufacture of such a machine (or types of machine) should be officially sponsored so that a group of unbiased persons with suitable qualifications could be formed to carry out such research and design.

10: Summary and conclusions

In attempting to reach any conclusions as to the likely future development or cost of heat pumps, it is difficult to find the correct criterion for judgement. What, for instance, is likely to be the value of the last one per cent of the world's oil or coal supplies? The last few per cent must inevitably be reached in the case of underground oil, probably within the lifetime of some readers of this book. Figure 33 shows a set of hypothetical curves in which the order of these values is indicated; we must be approaching the asymptotic values of cost in the case of oil or coal. Obviously, the value of an undeniable fuel-conservation machine such as the heat pump must increase in some ratio to the increase in fuel cost. In the period immediately after the Second World War two executives of the Edison Electric Co. in the USA and a group of people consisting of the B.E.I.R.A committee in Britain began to urge the need of the heat pump as a fuel-saver, at a time when the number of machines in use in each country was negligible. The number of machines now in use in the USA is numbered as being above one million with annual sales of 140 thousand - in Britain the number is still negligible, probably of the order of 100.

As technical progress proceeds in the art of obtaining low-grade heat cheaply, combined with the very considerable improvement possible in refrigeration compressor design and in refrigerants, the present Performance Efficiency Ratio of 2 to 3 will be increased to 4. This would mean that, for each electrically driven heat pump installed to provide a given amount of central house heating,

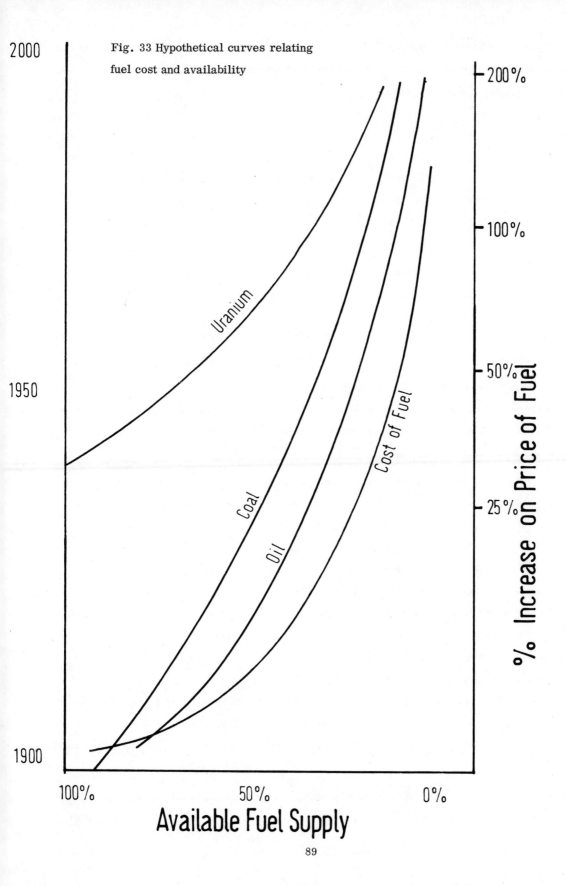

Fig. 33 Hypothetical curves relating fuel cost and availability

in place of resistance heating, the C.E.G.B. would save from one-half to three-quarters of the fuel used at present to provide that supply. The case can be shown quantitatively by assuming that one million consumers has all-electric heating, taking 120×10^6 Btu (35.17×10^6W) per annum, i.e. 35 000 kWh per year, with a demand of 13 kW. If it is further assumed that the supply authorities burn 1 lb (0.454 kg) of coal and 0.7 lb (0.318 kg) of oil per kilowatt hour, the fuel consumption per annum will be 16 million tons of coal or 12 million tons of oil. Even with a value of $R_\eta = 2$, the substitution of a heat pump in each case would involve annual savings of 8 and 6 million tons of coal and oil respectively, as a minimum. In time, as fuel grows more scarce and expensive and heat pump technology improves, the annual savings shown above could be increased by a factor of 1.5 to 2.0.

The use by the supply authorities of preset maximum demand load cutouts, to safeguard against a resistance heating load of high demand being incorporated with the heat pump would not be a new feature in electricity supply history; this would safeguard both the C.E.G.B. and the Electricity Boards. Their fear of loss of revenue, which has hitherto been the chief cause of lack of progress in heat pump development in Britain (vis-a-vis the USA) has always been groundless. A house-holder using oil fuel for central heating, instead of electricity, in looking for a remedy for heavily increased costs, can turn only to the heat pump as his remedy. In urban areas the only solution at present is the electrically driven heat pump. If this premise was true in 1950, it is trebly true in 1975.

It is unlikely, for a time and until the growing world shortages of fuel have made their full impact, that oil-engine driven heat pumps will be used for domestic house heating in urban areas. One development that may be of great influence in future technical progress is the Stirling cycle engine developed by Philips at Eindhoven, the Netherlands. A diagrammatic view of the engine is shown in Fig. 34. The feature of this engine lies in the regenerator which stores and restores heat directly, as heat, thereby effecting the change from t_1 to t_2, and from t_2 to t_1 instead of the two adiabatic operations in the Carnot cycle. An engine working, so as to do work, on the practical Stirling cycle has a higher efficiency than any engine working on an alternative type of practical cycle. Conversely, a heat pump working on the Stirling cycle will have a higher P.E.R. and a higher value of R_η than an engine working on any other cycle.

It is understood that efficiencies exceeding 95 per cent have been achieved by the regenerator. If this is so, it would appear possible that a Philips Sterling engine driving a similar machine as a heat pump would give a value of R_η exceeding 4, but with a somewhat lower value of q_2 recycled from the engine waste heat.

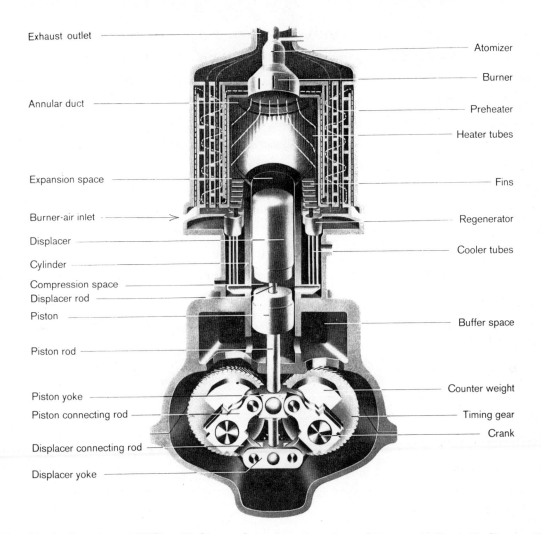

Exhaust outlet

Annular duct

Expansion space

Burner-air inlet

Displacer

Cylinder

Compression space

Displacer rod

Piston

Piston rod

Piston yoke

Piston connecting rod

Displacer connecting rod

Displacer yoke

Atomizer

Burner

Preheater

Heater tubes

Fins

Regenerator

Cooler tubes

Buffer space

Counter weight

Timing gear

Crank

Fig. 34. Section view of Philips Stirling cycle engine. (courtesy of Messrs Philips, Eindhoven, Holland)

Capital and Running Costs

Present-day costs for a heat pump supplied by the refrigeration industry for a heat output q_1 of 38,000 Btu/h (11,140 W) at £1600 has no true relation to a mass-production cost. An analogy can be made between a heat pump and a small motorcar costing £1000. The compressor is cheaper to make, and more simple than the motorcar engine and an electric-motor driven hermetic compressor no more expensive than the engine and gearbox. The two remaining major components of the heat pump, i.e. the evaporator and condenser, are of very little more complexity and cost than two good radiators or silencers. This would suggest that, if mas-production can provide the equivalent of the quarter of a car for £250, then mass-production of heat pumps should be undertaken and produced at that price. A recent decision to use public money to encourage the manufacture of electric heaters with 33 per cent fuel efficiency should be counterbalanced by a decision to support the manufacture of heat pumps with an overall fuel efficiency of 100 per cent.

For providing heat at the right temperatures required for space heating it can safely be asserted that no known device can provide the fuel efficiency that is provided by a heat pump driven by a reasonably efficient fuel-consuming engine. Such an engine, with an efficiency of $33\frac{1}{3}$ per cent driving a heat pump with a P.E.R. of 4 will return more than 100 per cent of the heat available in the fuel consumed and largely eliminate the ground or air coil. The development of fuel-saving devices such as those inherent in the Philips Stirling engine could lead to engine efficiency values of 0.4 (40%) and a value of $R_\eta = 4$, in which case, for each unit quantity of fuel consumed, we should obtain an amount of heat 1.4 times greater than the heat available in the fuel; the use of the term 'heat-multiplier' in 1852 by Lord Kelvin would then be vindicated. It is interesting to quote here a table (Table 13) prepared by the late Professor S.J.Davies, and which appeared in his "Heat Pumps and Thermal Compressors" (1950). This referred to a building in the southern counties which had a swimming pool used in summer and an assembly or dance hall used in winter. The table shows the relative quantities of fuel which would require to be burned for this particular use.

Table 13. Comparison between fuel consumption of coke-fired boilers and heat pumps, electrically and oil-engine driven.

Installation	Capacity	Mean fuel consump- tion, ton per week	Thermal efficiency Heat delivered	Annual fuel con- sumption
(a) Boilers, coke-fired	12.5 therm per h	(Coke)		(Coke)
Winter		1.67	50 per cent	90 tons
Summer		1.82	50 per cent	
(b) Heat pumps, driven by electric motors	h.p. of motors, 141	(Coal)		(Coal at station)
P.E.R., 3.5 Winter		0.88	87 per cent	39 tons
P.E.R., 6 Summer		0.56	150 per cent	
(c) Heat pumps, driven by oil engines	h.p. of engines, 99	(Diesel oil)		(Diesel oil)
P.E.R., 3.5 Winter		0.36	155 per cent	17 tons
P.E.R., 6 Summer		0.27	228 per cent	

The advantage which the electrically driven heat pump has gained over recent years, due to the much greater increases in the cost of solid fuels than for electricity, is shown in Tables 14 and 15. The first table relates to known and recorded costs in 1948 on an automatic stoker-fired coal boiler operated from 1940-45 and for the 60 kW heat pump which was substituted from 1945-50, known as the Norwich Heat Pump. The second table represents arbitrary modern capital and annual costs obtained by increasing the 1948 costs by

Chrisco Ltd.

Construction, Commissioning Engineering

Inspection Services

TREGER HOUSE
CIRCULAR ROAD
DOUGLAS
ISLE OF MAN
Telephone 0624 21546

TO ....

....

....19......

Table 14.   Relative 1948 costs of heating a building in Norwich by coal
firing and heat pump respectively.

Annual Operating Data for Comparison (1948)

Heat supplied to building during heating season	20 000 therm
Peak heat demand	8 therm
Average heating demand	5 therm
Calorific value of coal (½-in washed nuts)	12 000 Btu/lb
Price of coal per ton	65s. 0d.
Cost of electricity:	
For loads on-peak	£4 per kVA plus 0.6d per kWh
For loads off-peak	0.6d per kWh
Average thermal efficiency of combustion	55 per cent
Average seasonal cost of attendance, removing ashes, filling hoppers, etc., on coal-fired boilers	£230
Averaged annual cost of maintenance and repair of coal-fired boilers	£150
Capital cost of coal-fired boilers	£1 500
Capital cost of heat pump	£3 000
Additional cost of thermal storage vessel for heat pump	£500

	Coal-fired boilers	Heat pump	
		Without thermal storage	With thermal storage
Capital cost	£1 500	£3 000	£3 500
Averaged seasonal reciprocal efficiency	—	4.00	4.00
Tariff for electrical energy	—	£4 per kVA demand; 0.6d per kWh	0.6d per kWh
Annual capital charges	£225 (15 per cent)	£210 (7 per cent)	£245 (7 per cent)
Cost of coal (or electricity)	£440	£601	£367
Annual cost of attendance	£230 (including coal and ash handling)	—	—
Repairs and maintenance	£150	£50	£50
Replenishment of refrigerant	—	£25	£25
Total	£1 045	£978	£769
Cost per therm	£0.052	£0.044	£0.038

*Journal Inst. Mech. Engrs, vol. 158, no. 1, June 1948. 'The Norwich Heat Pump'.

Table 15.   Relative 1975 costs of heating a building in Norwich by coal firing and heat pump respectively.

Annual Operating Data for Comparison (1975)

Heat supplied to building during heating season	20 000 therm
Peak heat demand	8 therm
Average heating demand	5 therm
Calcrific value of coal (½ in washed nuts)	12 000 Btu/lb
Price of coal per ton	£26
Cost of electricity:	
For loads on-peak	£10.63 per kVA plus 1.13p per kWh
For loads off-peak	1.13p per kWh
Average thermal efficiency of combustion	55 per cent
Average seasonal cost of attendance, removing ashes, filling hoppers, etc., on coal-fired boilers	£920
Averaged annual cost of maintenance and repair of coal-fired boilers	£600
Capital cost of coal-fired boilers	£6 000
Capital cost of heat pump	£12 000
Additional cost of thermal storage vessel for heat pump	£2 000

	Coal-fired boilers	Heat pump	
		Without thermal storage	With thermal storage
Capital cost	£6 000	£12 000	£14 000
Averaged season reciprocal efficiency	—	4.00	4.00
Tariff for electrical energy	—	£10.63 per kVA demand; 1.13p per kWh	1.13p per kWh
Annual capital charges	£900 (15 per cent)	£8.40 (7 per cent)	(7 per cent)
Cost of coal (or electricity)	£3 520	£2 400	£1 770
Annual cost of attendance	£920 (including coal and ash handling)	—	—
Repairs and maintenance	£600	£200	£200
Replenishment of refrigerant	—	£100	£100
Total	£15 940	£3 550	£3 050
Cost per therm	£0.297	£0.178	£0.153

a factor of 4. The costs of coal and electricity which have been substituted are those current in Norwich in March 1975.

It will be noted that the 1948 costs per therm of £0.044 and £0.038, which were slightly in favour of the heat pump compared to the coal-fired cost of £0.053, have now become much more favourable to the heat pump in 1975, i.e. £0.178 and £0.153 compared with £0.297 for the coal-fired boiler. Reference of these latter figures to the curves in Fig. 4 show a reasonable coincidence if capital and annual costs were increased by a factor of 2 instead of 4.

These facts illustrate a very interesting point. What has been termed the use of 'raw' electricity, i.e. electric resistance heating, is only competitive with coal, gas or oil on a purely monetary basis if these fuels are burned at less than 40 per cent efficiency (see Fig. 4). But, on a fuel conservation basis, 'raw' electricity can only compete if the alternative fuels were being burned at a lower rate of efficiency of the order of $33\frac{1}{3}$ per cent which corresponds to the average distribution heat efficiency of electrical energy likely to be reached in this decade. It is true that any of the alternative fuels if used normally at, say, 50 per cent efficiency, can have this value increased to 150 per cent if the particular fuel is used to drive a heat pump with a value of $R_\eta = 3$. However, the use of an electrically driven heat pump would result in a fuel efficiency of 100 per cent, an efficiency which none of the other fuels can reach in practice.

That some definite guidance and control of the use of fuel is imminent seems to me to be certain. Panics cannot be allowed to occur repeatedly without some indication that their causes are being investigated. The phrase 'energy crisis' is used often without the words being understood. A 'crisis' is defined as a time of danger and a turning point, yet the imminence of the crisis has appeared to be ignored (or not appreciated) by all governments of the last two decades. If, as I believe, the heat pump can provide a given amount of heat with a lower fuel consumption than any other method of fossil-fuel burning, no sensible government can continue to ignore this method. It is fashionable (but not necessarily true) to suggest that what private enterprise has failed to do can be done by government. Chapter 2 might suggest that heat pump development is a case in point. It is certain that this development must not be left to the ploy of the many interests to whom fuel conservation is secondary to monetary economy.

It is not usually appreciated that coal (and most other solid fuels) has increased in price at least eighty times- from £0.5 to £40 per ton in the case of coal - since the beginning of the century, whereas electricity has actually decreased in price during the same period. Up to 1920, approximately, electricity was sold at or near, the statutory maximum flat rate which varied from 2.5p and 5p per kWh until two part tariffs were introduced in 1920. In 1945 off-peak electricity was being sold at 7 kWh per 1975 penny plus a small standing charge. We spend a great deal of time considering whether we can increase the efficiency of our power stations from 38 to 40 per cent without realising that electricity delivered at $33\frac{1}{3}$ per cent efficiency to be used in driving a heat pump represents a power station efficiency of fuel use of 100 per cent.

This book deals only with one small facet of the country's fuel consumption. There are other much

95

more fruitful fields than the domestic field in which the heat pump could provide fuel conservation, some of which are shown in Table 16.

Table 16. Relative efficiency of various types of heating by direct coal burning and heat pump respectively.

Operation	Average maximum temp. of operation (°F)	P.E.R. of heat pump with low-grade heat at 40°F	Fuel consumption per million Btu	
			Direct coal burning: (assuming 60% efficiency for boiler (lb. of coal at boiler)	With heat pump (lb of coal)
Building heating:				
(i) with low-temperature radiating surfaces	120	3 to 3.5	160	50
(ii) with circulated air	95	4.5 to 5.5	160	32
Domestic water heating	130	2.5 to 3	160	60
Swimming bath water heating	75	8	160	20
Electric train heating	130	3	160	53
Milk pasteurising (heating from refrigeration)	?	?	160	56
Pit-head baths	110	4	160	40
Cinema heating	95	—	160	40
Heating of dye vats using thermocompressor	212	2	160	80

As mentioned earlier in the book, for each unit of fuel burned in the normal way, at least one half of the generated heat serves no purpose and is wastefully dissipated into the atmosphere. Yet where a heat pump is in operation, all, and sometimes more than all, the heat in the fuel is used. The cost of fuel burned in 1974 in Britain was £6000 million. Therefore the value of the wasted heat lost to the atmosphere was at least £3000 million. If only 10 per cent of the wasted heat could be saved by the use of heat pumps it would involve a saving of £300 million on the annual national expenditure.

Let us look at this from another angle. If £3000 million is wasted annually due to the 50 per cent wastage, what are the economics of using a heat pump?

I realise that the heat pump is not suitable in all heating situations, though I believe it could be used in most cases. However, let us imagine its use in only 1 per cent of all situations. Normal business practice justifies a capital expenditure of at least £3.00 for each £1.00 of continuous annual return. As such, 1 per cent saving on £3000 million would save £30 million per year, thus justifying capital expenditure of £90 million. I would remind the reader that the '3 to 1' ratio, and more probably the '1 per cent' referred to are conservative estimates. In the case of North Sea oil a capital expenditure of £25 x 10^9 million is held to be justified in order to give an annual revenue of £3 x 10^9 million, i.e. 8 per cent on investment.

Mention may be made at this point as to what appears to be the lack of direction of present research work. There is a considerable amount of test data available for domestic heat pump development, e.g. the Morris Motor Co's prototype work, the two test houses and machines built by Messrs Lucas, the B.E.I.R.A. series of tests and reports and, finally, my test results embodied in this book. Yet, so far as can be ascertained, the Capenhurst tests carried out by the Electricity Council ignore these data and are limited to tests on an undersized imported heat pump of unsuitable design. Gresham's Law can apply equally to heat pump tests as to money. Instead of spending effort on testing an unsuitable machine, under unsuitable conditions to demonstrate how badly such a machine can work, it would be preferable to commence with a target, say of a minimum P.E.R. = 3 and to design a new machine that will achieve the criterion set for it. Such is the purpose of this book.

The Cost Effectiveness of a Heat Pump

Modern methods of costing require a setting of two parameters - cost and effectiveness. If two alternative schemes or methods were found to require the same annual expenditure, the scheme which has the greater 'effectiveness' in serving the community would be preferred. There would be no hesitation in government or private business making a choice if one scheme cost one-third of the alternative scheme and was twice as effective.

Final Conclusions

(1)   This book explains the principles of construction of a fuel-saving machine (a heat pump) suitable for centrally heating, inter alia, domestic dwellings. The principles of the machine have been known since 1852.

(2)   The machine, if properly designed, uses only one-half to one-third of the fuel used by any other method of fuel-burning when providing a given amount of heat. The saving in the annual cost of fuel purchased by the consumer is of the same order, i.e. 50 to 70 per cent for a given heat output.

(3)   One such machine as described in this book has been in use during each winter heating season since 1952. The machine is part of a fully instrumented test bed and continuous records have been kept which prove that the annual cost of heating the house is one-third of the cost of using electricity in the conventional way.

(4)   It is estimated that the annual cost of the heat lost from the fuel burned in Britain is £3000 million. The heat pump could be used to save at least 25 per cent of this heat loss, thereby providing a saving of 25 per cent of the fuel used nationally, and could provide a saving in consumers' fuel bills of £750 million.

(5)   There is historical evidence that there has been a lack of enthusiasm regarding development of this machine by the Electricity Boards and the Electricity Council. There is no reason to think that any greater enthusiasm is likely to be shown by the Coal Board or by manufacturers of existing fuel-

burning or heat consuming appliances. Development of this, or any other machine of such a radical nature, is bound to evoke opposition from interests vested in fuel consumption.

(6)   The most suitable body to supervise and encourage development of the heat pump is a national body such as the National Enterprise Board.

(7)   The machine consists of four major components each simpler but closely related to components at present being mass-produced by a declining motorcar industry. These components consist of a 3- or 4-cylinder 900 c.c. compressor, two heat exchangers and a metering valve. The present manufacturing plant could be used, almost without change, to manufacture heat pumps.

(8)   The 1975 estimated cost of mass-producing a 4 kW domestic heat pump should not exceed £250 and the machine could be sold at £500, which is comparable with oil and coal central-heating plant costs.

Appendix A   Analysis of thermodynamic quantities in heat pump cycle

General Principle of Operation of the Heat Pump

In any heating process the heat used must be at a temperature, say, t_1, higher than that of the surroundings, say, t_2. Instead of using a process wherein a chemical reaction, such as the combustion of fuel, takes place accompanied by the evolution of heat at a high temperature (with subsequent degradation of heat), the process of upgrading heat, as with a heat pump, may be used. The latter process is based on the property of substances, e.g. gases and vapours, becoming hotter when compressed, due to the work of compression being changed into internal energy in the substance.

If a gas is compressed adiabatically from p_2 to p_1, the temperature will increase from t_2 to t_1 in the ratio of

$$\frac{t_1}{t_2} = \frac{p_1}{p_2} (Y - \frac{1}{Y}) \ .$$

When a gas or vapour is compressed, as in the heat pump, mechanical work W must be supplied, as will be shown. The mechanical work thus supplied is transformed into energy in the gas or vapour compressed, this being the real cause of heating. The work could be used directly to produce heat, e.g. by friction between objects or by producing turbulent motion in a gas or liquid. The method used in the heat pump process to compress gases or vapours so as to produce heat at a higher temperature is, therefore, what might be termed a mechanical 'generation' of heat as compared with the chemical generation of heat and subsequent fall in temperature occurring during combustion.

As stated, if, instead of using a heat pump, mechanical work were to be used directly to generate heat, the generation of 1 Btu requires an expenditure of 778 ft lb of mechanical energy or, alternatively, 1 Btu of heat is obtained at the expense of 1 Btu of energy. By using a heat pump to perform mechanical compression of gases or vapours, it is obviously impossible to obtain more than 1 Btu of heat from 1 Btu of mechanical work. But it will be shown that existing quantities of heat such as heat contained in the atmosphere or in the earth at atmospheric temperature, say, 50°F, can be brought to a higher temperature, say, 80°F, after compression, by addition of the mechanical work of compression and therefore of internal energy. Thus, 778 ft lb of mechanical energy will produce

1 Btu of heat which augments an existing quantity of heat whatever the temperature of that existing heat might be. Hence, if 1 lb of water exists at 50°F, 30 x 778 ft lb of mechanical energy can be added as 30 Btu of heat, raising the temperature of the water to 80°F. Alternatively, if the water was at 80°F initially, the addition of the same quantity of heat in the form of mechanical energy would raise the temperature to 110°F without the degrading of thermal energy inherent in, say, a combustion process. Thus, an existing quantity of heat can be brought to a higher temperature by the expenditure of mechanical work, and it will be shown that this expenditure increases with the temperature difference sought.

The process of the heat pump and the manner in which a relatively great heat effect can be obtained from a relatively small expenditure of mechanical or electrical energy, are now shown in a general manner by considering first a compression and then an expansion operation, comprising a cycle of operations for a unit mass of 1 lb of air. It will be assumed that all operations are reversible and that when the cycle is completed, the working substance has been returned to its initial condition.

Assume that 1 lb of ambient air at atmospheric pressure (14.7 p.s.i.a.) and 50°F, is compressed adiabatically in a compression cylinder until its temperature rises to 80°F, when the pressure will have increased to approximately 17.9 p.s.i.a. The air is then delivered into an ideal receiver such that the air can give up heat to the atmosphere and change its volume at constant pressure:

$$\text{Let } t_2 = 50^{\circ}F = 510^{\circ}R \ (10^{\circ}C)$$
$$\text{and } t_1 = 80^{\circ}F = 540^{\circ}R \ (27^{\circ}C)$$

(these are the temperatures used by Kelvin in his original paper, describing the operation of his 'warming machine'.)

Then the following amount of heat q_1 will have been given up to the atmosphere surrounding the receiver:

$$q_1 = C_p \ (80^{\circ}F - 50^{\circ}F) = C_p \ (540^{\circ}R - 510^{\circ}R) = C_p \ (t_1 - t_2),$$

i.e. $\quad q_1 = 0.24 \ \text{x} \ 30 = 7.2 \ \text{Btu}.$

The work of compression W_c will have been:

$$W_c = JC_p \ (540-510) = 5601.6 \ \text{ft lb}$$

or $\quad \dfrac{5601.6}{J} = 7.2 \ \text{Btu}.$

Thus, to obtain the 7.2 Btu of heat from the compressed air, an equivalent amount of mechanical work $= \dfrac{5601.6 \ \text{ft lb}}{778} = 7.2 \ \text{Btu}$ had to be expended.

But the air in the receiver, after giving up its heat $= q_1$ or 7.2 Btu, isothermally, remains at a pressure of 17.9 p.s.i.a. and therefore has an excess of pressure remaining $= 17.9 - 14.7$ or

3.2 p.s.i.a. above its original atmospheric pressure. The air is still, therefore, capable of doing mechanical work, down to 14.7 p.s.i.a. in this ideal machine, after giving up the heat q_1 added during compression. Let this potential work be used in an expansion cylinder coupled, without loss, to operate the compression cylinder so as to do work = W_e.

Then the net work to perform the compression operation will be reduced from W_c to W_n, where:

$$W_n = W_c - W_e.$$

W_e is less than W_c because of the reduction in volume of the air after cooling.

Thus, with an expansion of only $W_n = W_c - W_e$ ft lb or $\dfrac{W_n}{J}$ Btu an amount of heat q_1 at 80°F will have been obtained. The ratio between the heat given up at 80°F (q_1) and the net work required to provide this heat $\dfrac{W_n}{J}$ will be:

$$\frac{q_1}{W_{n}/J} = \frac{W_c}{W_c - W_e} = \frac{1}{1 - W_e/W_c}.$$

Now, since W_e is less than W_c, the value of $\dfrac{q_1}{W_{n}/J}$ will also represent the ratio of the quantity of heat q_1 made available at the higher temperature t_1 per Btu to the mechanical work (net) done by the motor driving the heat pump. As will be indicated later, the value of q_1 and W_n for the conditions stated above are:

$$\frac{q_1}{W_{n}/J} = \frac{W_c}{W_c - W_e} = \frac{18}{1}.$$

Considering the two operations, the work available during expansion (in the expansion cylinder) is less than the work which has to be done during the operation in the compression cylinder. Both operations have similar pressure limits, but the volume of the compressed, and cooled, air entering the expansion cylinder is less than the volume at the end of compression in the compression cylinder. The ratio of volume will be:

$$\frac{v_a}{v_d} = \frac{\text{volume entering expansion cylinder}}{\text{volume leaving compression cylinder}} = \frac{t_2}{t_1}.$$

It was stated that:

$$\frac{W_c}{J} = C_p (t_1 - t_2) = q_1. \qquad . \qquad . \qquad . \qquad . \qquad (1)$$

Hence
$$\frac{W_e}{J} = C_p (t_1 - t_2)\frac{t_2}{t_1} = C_p t_2 \left(1 - \frac{t_2}{t_1}\right).$$

Thus
$$\frac{W_n}{J} = W_c - W_e = C_p (t_1 - t_2)\left(1 - \frac{t_2}{t_1}\right). \qquad . \qquad . \qquad (2)$$

Again, since $\dfrac{q_1}{W_{n}/J} = \dfrac{W_c}{W_c - W_e}.$

Then
$$\frac{q_1}{W_{n/J}} = \frac{t_1}{t_2}.$$

$$\text{(where denominator } (1 - \frac{t_2}{t_1})) \tag{3}$$

Considering still further the process in the expansion cylinder: in view of the adiabatic expansion in the expansion cylinder, a drop in temperature of the air occurs from t_2 to t_3 ($^{\circ}$R). This occurs for the following reasons.

The driving work available at the expansion motor amounts to $\frac{W_e}{J} = C_p(t_2 - t_3)$ Btu, but for the same work it was shown above that:

$$\frac{W_e}{J} = C_p \frac{t_2}{t_1} (t_1 - t_2). \tag{4}$$

Equating these two quantities gives:

$$t_2 - t_3 = \frac{t_2}{t_1}(t_1 - t_2) \text{ and } t_3 = \frac{(t_2)^2}{t_1} = \frac{510^2}{540}$$

or 481.35°R, i.e. 21.60°F, for the temperature values chosen. Therefore, in the expansion cylinder, the temperature drops below the original atmospheric temperature.

Therefore, at this stage of the cycle, we now have the working substance at a temperature lower than its initial temperature but restored to the initial pressure. To complete the cycle, therefore, air must take up heat $= q_2 = C_p (t_2 - t_3)$ before the cycle can be completed, and this heat must be taken up from the surrounding atmosphere. But:

$$t_3 = \frac{t_2^2}{t_1} .$$

Therefore $\quad q_2 = C_p \frac{t_2}{t_1} (t_1 - t_2)$.

Now $\quad \dfrac{W_n}{J} = C_p (t_1 - t_2) (1 - \dfrac{t_2}{t_1})$ as shown earlier.

And $\quad \dfrac{q_2}{W_{n/J}} = \dfrac{C_p \dfrac{t_2}{t_1}(t_1 - t_2)}{C_p (t_1 - t_2)(1 - \dfrac{t_2}{t_1})} = \dfrac{t_2}{t_1 - t_2} . \tag{5}$

It is possible, however, to exhaust the air at t_3 from the expansion cylinder to the atmosphere and to take into the compression cylinder a new charge of air at t_2 directly from the atmosphere which would contain the quantity of heat $= q_2$. In this case the exhaust air would become heated by the atmosphere from t_3 to t_2 and an equal mass of air already heated to t_2 would have been taken into the compression cylinder.

102

Considering the complete cyclic operation, the following data have been established:

From (3) $\dfrac{\text{Heat given up at } t_1}{\text{Heat equivalent of work input}} = \dfrac{q_1}{W_{n/J}} = \dfrac{t_1}{t_1 - t_2}$.

From (5) $\dfrac{\text{Heat taken in at } t_2}{\text{Heat equivalent of work input}} = \dfrac{q_2}{W_{n/J}} = \dfrac{t_2}{t_1 - t_2}$.

Now $\dfrac{t_1}{t_1 - t_2} = 1 + \dfrac{t_2}{t_1 - t_2}$.

Therefore $\dfrac{q_1}{W_{n/J}} = 1 + \dfrac{q_2}{W_{n/J}}$.   .   .   .   .   (6)

Multiplying each side by $W_{n/J}$ gives

$$ q_1 = W_{n/J} + q_2 . \qquad . \qquad . \qquad . \qquad . \qquad (7) $$

We can, therefore, say that the quantity of heat delivered $= q_1$ is greater than the heat equivalent of the mechanical work done during the whole operation, by an amount $= q_2$. If it is assumed (in expressions (6) and (7)) that $W_{n/J} = 1$ Btu, it can be said that, for an expenditure of 1 Btu of work, an excess heat $= q_2$ has been gained.

But since energy cannot be created, this excess heat must have been available within the cycle at the beginning of the operation, i.e. it must have been present in the ambient air taken from the compression cylinder before compression. Thus, by expending $1 \times J$ ft lb or the heat equivalent of 1 Btu of mechanical work and adding this to q_2, the original amount of heat (q_2 at 50°F) has been raised to 80°F.

That this is so may be shown by considering the operation in terms of the change of internal energy within the system.

The 1 lb (unit mass) of air before isentropic compression was at a pressure p_2 of 14.7 p.s.i.a. and had a volume V_2 of 12.842 ft^3, with an amount of internal energy E_2 corresponding to the atmospheric temperature of 510°R, where 510°R $= 460^{\circ}$F $+ 50^{\circ}$F on the scale of Absolute Temperature.

The total heat H_2 of the gas before compression can be assumed as being H_2 Btu/lb, when:

$$ H_2 = C_p t_2 = C_v t_2 = \frac{p_2 V_2}{J} = E_2 + \frac{\text{Flow work}}{J} = q_2 E_2 \left(H_2 - \frac{\text{Flow work}}{J} \right) . $$

After isentropic compression the temperature of the air is raised by 30 degF to 540°R.

Hence, the increase in internal energy, i.e.:

$$ E_1 - E_2 = C_v (t_1 - t_2) = 0.1715 \times 30 = 5.145 \text{ Btu/lb.} $$

Increase in flow work $= (R/J)(t_1 - t_2) = \dfrac{53.18}{778} \times 30 = 2.0504$ Btu/lb.

Giving $H_1 - H_2 = C_p (t_1 - t_2) = (E_1 - E_2) + R/J (t_1 - t_2) = 5.145 + 2.0504 = 7.1954$ Btu/lb.

i.e. $q_1 = (E_2 + 5.145) + (p_2 V_{2/J} + 2.0504)$

or $q_1 = (E_2 + p_2 V_{2/J}) + 7.1954$ Btu/lb.

But $(E_2 + p_2 V_{2/J}) = q_2$.

Therefore

$q_1 = q_2 + 7.1954 = q_2 + W_{C/J}$ (at this stage in the cycle).

At this stage, therefore, as was shown earlier, there has been added to an existing amount of heat $= q_2 = H_2$, an amount of heat equal to the heat equivalent of the work of compression, i.e 7.1954 Btu/lb.

During the second stage of the operation an amount of heat is abstracted from the gas = $C_p (t_1 - t_2)$ which can be shown to be equal to the heat equivalent of the work done during isentropic compression. The working substance thereby falls to $t_2 = 510°R$ but the pressure (ideally) remains constant at p_1. Expansion then occurs until the temperature of the working substance has fallen from $510°R$ to $481.6°R$ and a corresponding amount of work will be done by the working substance. The value of E will fall as the temperature decreases, as will the inherent flow work value. The total amount of energy given up $(W_{e/J})$ is:

(a) In flow work of expansion $= \dfrac{p_e V_e - p_f V_f}{J} = R_{/J} (510 - 481.7) = 1.936$ Btu/lb.

(b) In internal energy $= C_V (510 - 481.7) = 4.853$ Btu/lb.

(a + b) = Heat equivalent of work of isentropic expansion = 6.789 Btu/lb.

Then $W_{n/J} = W_c - W_{e/J} = 7.1954 - 6.79 = 0.4054$ Btu/lb.

The working substance is now at a temperature $t_3 = 481.6°R$. In order that the ideal cycle may be completed and the working substance return to its original condition, i.e. $t_2 = 510°R$, the following quantities of heat energy must be taken in from the atmosphere:

(a) As flow work $- \dfrac{R}{J} (t_2 - t_3) = \dfrac{R}{J} (510 - 481.6) = 1.918$ Btu/lb.

(b) As internal energy $- C_V (t_2 - t_3) = 0.1715 (510 - 481.6) = 4.871$ Btu/lb.

(a + b) = $H_2 = q_2$.

Thus it can be shown that the working substance commenced the cycle with a value of $H_2 = 6.789$ Btu/lb. By adding a net amount of work, with a heat equivalent of o.4054 Btu to each pound of working substance, an amount of heat = $H_1 = 7.1944$ Btu/lb was made available at $t_1 = 540°R$.

Then, for the whole cycle:

$$q_1 = H_1 = q_2 + W_{n/J}$$

i.e.   7.1954 = 6.789 + 0.4054 Btu/lb.

i.e by expending an amount of energy = 0.4054 Btu/lb and adding this to an existing quantity of heat = 6.789 Btu/lb at 510°R, an amount of heat = 7.1954 Btu/lb became available at 540°R. This justifies the equation on page 101 that:

$$\frac{q_1}{W_{n/J}} = \frac{18}{1} \text{ (approx.).}$$

The careful reader will doubtless have noticed a discrepancy in the earlier calculations. Reference has been made to an ideal heat pump (air type) working between 80°F (26.7°C) and 50°F (10°C) in which the work input is '1/36th of energy of the heat communicated'. Yet, on page   reference is made to an ideal heat pump working on the Carnot cycle, between the same temperatures in which the work input is only '1/18th of the energy of the heat communicated'.

The heat pump postulated by Lord Kelvin was described in detail in his two papers, to which reference should be made for a more complete description and consideration of the mathematical analysis which is given. It worked on a different cycle from that shown in Fig. 5b; his cycle was sub-atmosphere. During the first operation (AB in Fig. 5b) air at atmospheric temperature and pressure was allowed to enter an 'ingress' cylinder (connected to an air receiver) when the piston was at the beginning of its stroke. Since the pressure in the cylinder behind the piston was below atmospheric pressure it was assumed that expansion work W_e would be done on the piston = $R \log_e r$. The cylinder and its associated air receiver was to be so designed that heat could be taken in from ambient sources at temperature t_2 so as to maintain the temperature of the air in the receiver at temperature t_2. Thus the operation (AB in Fig. 3b) would, hopefully, be an isothermal operation doing work. The next step was to draw air from the receiver at t_2 and to compress it adiabatically in the opposing 'ingress' cylinder from t_2 to t_1. So far then, the operations Ab and BC of the Carnot cycle (Fig. 3d) have been followed but at this point the Kelvin cycle departs from the Carnot cycle.

The third, and final, operation is to expel the air at t_1 from the 'ingress' cylinder directly into the room to be heated thus giving out an amount of heat $q_1 = C_p (t_1 - t_2)$, i.e. the cycle follows the path ABCA in Fig. 5b instead of path ABCDA. Less heat is given out at t_1 in the Kelvin cycle and the operation CA is irreversible. But the reason for the discrepancy mentioned above may be found interesting to a serious reader and will provide him with a new problem in thermodynamic theory.

Appendix B   Calculating cycle values

The values which are given in Fig. 35 and which are later used in exercises are indicative only. Correct values for a given case should be found from standard tables showing enthalpy, entropy and other values as issued by refrigerant manufacturers.

Design should start by graphically reproducing the four elements of the vapour compression (i.e. Rankine) cycle on a chart (see figs. 7a and 16) on which the scale of quantities being considered are shown as direct scalar values. Of the several types of charts possible the pressure/enthalpy or p/h chart is the most suitable and most generally used. Charts may be obtained commercially with printed scalar values for each of the refrigerants in common use. The line LL is the 'saturated liquid' line (Fig. 35) which represents state

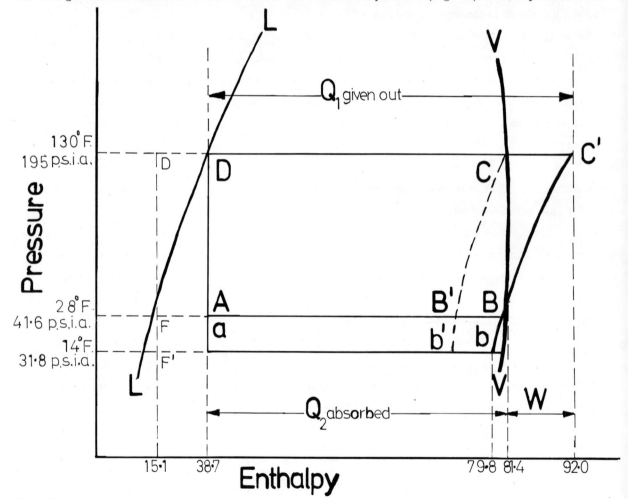

Fig. 35. P/h curves for two cycles (case (a) and case (b)).

points where the liquid has a temperature corresponding to the saturated pressure. The line VV is the 'saturated vapour' line and is the locus of state points which represent vapour having a temperature corresponding to a given saturation pressure. In the area to the left of line LL is the 'sub-cooled region' in which the temperature of the liquid refrigerant is less than the saturation temperature corresponding to the pressure of the liquid. In the 'wet area' between lines LL and VV a mixture of vapour and liquid exists. The area to the right of the line VV is the 'superheated region' in which the temperature of the vapour is greater than the saturation temperature corresponding to the pressure of the vapour. The 'state' of the refrigerant at any point on the chart is determined by any two properties on the chart except in the wet area when the 'quality' must be given in addition to the other two properties.

In Chapter 7 an outline design was prepared for two heat pumps based on the use of p/h charts. In this appendix the method of constructing ideal cycles from p/h charts is more fully explained and the factors which affect conversion from an ideal to a practical heat pump are considered.

As an exercise two separate cycles are drawn in Fig. 35 similar to the two heat pumps outlined in Chapter 7.

(a) Cycle using a ground coil (Cycle ABCD)

Antifreeze enters the evaporator at 33°F and leaves at 21°F, with a refrigerant pressure of 37.15 p.s.i.a. (2.152 bar) and an equivalent temperature of 22°F. Compression is assumed to commence at the point where the horizontal line AB intersects the saturated vapour line at B. Compression then takes place in the superheat region to the right of line VV (dry compression) and is assumed to follow a 'constant entropy' line as shown on the p/h chart. A final pressure of 195 p.s.i.a. (13.2 bar) is reached after compression with a corresponding temperature of 130°F. The condenser must give out 38,000 Btu/h.

(b) Cycle using ambient air (Cycle abCD)

Air is assumed to enter the evaporator at 33°F and to leave at 18°F and to give a refrigerant pressure of 31.8 p.s.i.a. with a corresponding refrigerant temperature of 14°F. Dry compression takes place as before to 195 p.s.i.a. (13.2 bar) and 130°F (cycle abCD) in order to provide heat at 130°F as for the ground coil machine. The condenser must give out 38,000 Btu/h.

The cycles ABCD and abCD provide the data to ascertain the following properties:

q_1	= Condenser duty (per 1 lb of refrigerant)
q_2	= Evaporator duty (per 1 lb of refrigerant)
q_R	= Weight of refrigerant to be circulated
I.H.P.	= Indicated horsepower
V_c	= Theoretical volumetric capacity of compressor
W_c	= Isentropic work of compression
R_η	= q_1/W

It should be noted that cycles abC'D and ABC'D represent ideal conditions (except that temperature fall at the heat exchangers has been allowed for). These ideal values may now be found by subtending each point to the horizontal ordinate (enthalpy) so as to find the heat value (per 1 lb of refrigerant) at each point. (Enthalpy = total heat per lb of refrigerant).

How to draw a cycle on the p/h chart

Draw a horizontal line between LL and VV at the three specified pressures. When the lower two lines meet line VV mark off state points b and B respectively. From points b and B follow the nearest 'constant entropy line' on the chart until each meets an extension of the upper horizontal line CD. This will give state point C'. Where the upper horizontal lines intersect line LL mark off state point D and draw a vertical line which intersects the lower horizontal lines; these lower intersections give state points a and A representing the beginning of each cycle. The values of total heat (enthalpy) for each state point are given by extending each point down to ordinate along the base of the diagram.

The distances between state points on the lower enthalpy scale represent the following heat values (per 1 lb of refrigerant circulating);

$$b - a \text{ and } B - A = q_2 = \text{heat taken in at evaporator}$$
$$C' - b \text{ and } C - B = w = \text{work/heat value of compression}$$
$$C' - D = q_1 = \text{heat given out at condenser}$$

Heat values taken from the p/h chart are as follows;

(a) Cycle ABC'D using ground coil

$$q_1 = \text{heat given out} = C' - D = 92 - 38.7 = 53.3 \text{ Btu/lb}$$
$$q_2 = \text{heat taken in} = B - A = 81.4 - 38.7 = 42.7 \text{ Btu/lb}$$
$$w = \text{work/heat input} = C' - B = 92 - 81.4 = 10.6 \text{ Btu/lb}$$
$$\text{sub-cooling} = D - E = 38.7 - 15.1 = 23.6 \text{ Btu/lb}$$

$$q_1 = 38,000 \text{ Btu/h}$$
$$q_2 = \frac{42.7}{53.3} \times q_1 = 30,400 \text{ Btu/h}$$
$$q_R = \text{mass flow of refrigerant} = \frac{30,400}{53.3} = 712 \text{ lbs/h} = 11.87 \text{ lbs/min}$$
$$\text{I.H.P.} = \frac{q_R \times w \text{ lbs } 778}{33,000} = \frac{11.87 \times 10.6}{42.42} = 2.97 \text{ I.H.P.}$$

Ideal piston displacement = 11.87 lbs/min x 1.081 ft^3 = 12.83 ft^3/min

$$\text{P.E.R. (Ideal)} = {q_1}/{w} = \frac{53.3}{10.6} = 5.03$$

$$p_1/p_2 = \frac{195}{37.15} = 5.25$$

Heat available from sub-cooling = 712 lbs x 23.6 = 16,800 Btu/h

(b) Cycle abC'D using air

$$q_1 = \text{heat given out} = C' - D = 92 - 38.7 = 53.3 \text{ Btu/lb}$$
$$q_2 = \text{heat taken in} = b - a = 79.8 - 38.7 = 41.1 \text{ Btu/lb}$$
$$w = \text{work/heat input} = C' - b = 92 - 79.8 = 12.2 \text{ Btu/lb}$$

q_1 = 38,000 Btu/h

q_2 = $\frac{41.1}{53.3}$ x 38,000 = 29,300 Btu/h

q_R = mass flow of refrigerant = $\frac{29,300}{41.1}$ = 713 lbs/h = 11.87 lbs/min

I.H.P. = $\frac{11.87 \text{ x } 12.2}{42.42}$ = 3.41 I.H.P.

Ideal piston displacement = 11.87 x 1.163 = 13.8 ft^3/min

P.E.R. = $\frac{q_1}{w}$ = $\frac{53.3}{12.2}$ = 4.37

p_1/p_2 = $\frac{195}{31.8}$ = 6.13

Table 17. Comparison of two ideal heat pumps.

	Case A Ground Coil	Case B Air
q_1 (Btu/h)	38,000	38,000
q_R (Btu/h)	713 lbs/h	713 lbs/h
I.H.P.	2.97	3.41
Piston displacement	12.58 ft^3/min	13.8 ft^3/min
R_η	5.03	4.37
Pressure ratio	5.25	6.13

It will be seen that the ideal characteristics for each machine are very similar but favourable to the ground coil machine. The most important practical difference will be that the ground coil machine has no icing problem to contend with and the temperature and suction pressure will remain constant over the complete heating season. The air machine has neither of these advantages. Provision must be made for de-icing over 90 per cent of the heating season also. If the compressor is designed with a piston displacement of 13.8 ft^3/min and full load at 32°F it will be displacing 15.6 ft^3/min when the air temperature rises to 45°F, with a consequential overload of 26 per cent.

On the assumption that, to be viable, a commercial heat pump must have a P.E.R. = 3 it may prove an interesting exercise to ascertain the pressure rates and other factors of design which would permit this value to be obtained. The considerable quantity of sensible heat available and the frictional, etc. heat from the compressor are important factors. The serious inefficiency of design of British compressors designed normally for refrigeration purposes is a further factor; the practical loss due to temperature fall in the heat exchangers has already been allowed for but improved heat exchanger design should permit an improvement in this important design factor and consequent decrease in the value of the pressure ratio.

The Effects of Sub-Cooling

A correspondent has written questioning the benefits of sub-cooling. The benefits can be demonstrated by reference to Fig. 17.

In Fig. 17 it can be proved that area aDA = area AA_1FH and also, that the work done, when an expansion valve is used in either the refrigeration or heat pump operation, is represented by area aBCDa. For the refrigeration operation we have :

$$C.O.P. = \frac{\text{area } BA_1FG}{\text{area aBCDa}} = \frac{\text{Heat absorbed}}{\text{Work done}}$$

Suppose that, by some method, e.g. by using an expansion cylinder, an amount of extra work = w can be done in the cylinder; or, alternatively, an additional amount of heat = Jw that would not be available otherwise can be used; w is represented by area aAD and AA_1FH in Fig. 17. Also, let the C.O.P. for the improved operation be Q_2/W. Then we have:

$$C.O.P. \text{ using improved cycle } = \frac{Q_2}{W} \qquad . \qquad . \qquad . \qquad . \qquad (1)$$

$$\text{and} \quad C.O.P. \text{ using expansion valve } = \frac{Q_2 - w}{W + w} \qquad . \qquad . \qquad . \qquad (2)$$

Obviously the value in (1) is greater than the value in (2)

Worked Example of Effect of Sub-Cooling

A heat pump on cycle ABCD of Fig. 16 is between 120°F and 20°F with a mass flow of 10 lbs/min (600 lbs/h) of F12 having a specific heat of liquid of 0.227 Btu/lb. A mass flow of antifreeze liquid of 2000 lbs/h is entering the evaporator at 32°F and is required to supply 25,000 Btu/h to maintain the refrigerant at 19.5°F as it leaves the evaporator. The antifreeze falls in temperature from 32°F to (32°F $- \frac{25,000}{2000}$) $= 19.5^{\circ}$F giving an average temperature of 25.75°F. Now let the heat pump work on ideal cycle FBCDF (Fig. 16) in which a quantity of heat = 600 lbs/hr x 0.227 (spec ht) x (120°F $-$ 20°F) or 16,620 Btu/h of sensible heat is taken from the condensate and given to the antifreeze entering the evaporator at 32°F. The effect will be, ideally, to raise the temperature of the incoming antifreeze from 32°F to (32 $+$ $\frac{16,620}{2000}$) $= 40.31^{\circ}$F.

But the antifreeze must now supply to the evaporator 25,000 + 16,620 = 41,620 Btu/h. If the flow of antifreeze could remain constant at 2000 lbs/h there will be a temperature drop $= \frac{41,600}{2000} = 20.8^{\circ}$F, i.e. from 42.42°F to 19,62°F giving an average temperature to 30.75°F instead of 25.75°F for cycle ABCD. The effect will be to decrease the value of the pressure ratio from P_1/p_4 to P_1/p_3 and thereby to increase the efficiency of the cycle by reducing the value of work done giving cycle EB_1CDE in Fig. 16.

Ideal Cycle for Engine Driven Heat Pump (Fig. 21)

Because of the higher temperatures and pressures involved the refrigerant F12 (CCl_2F_2) is unsuitable and

F21 ($CHFCl_2$) will be used, with a ground coil. The antifreeze liquid circulating in the ground coil is assumed to leave at $34°F$, at the rate of 2000 lbs/h.

This incoming liquid at $34°F$ ($1.1°C$) will first absorb an amount of heat = 9240 Btu/h (2707 W) (see page 75) from the engine cooling water and the temperature will rise by $4.1°F$ to $38.1°F$ ($3.4°C$). Next will be the heat from the engine exhaust of 13,510 Btu/h which will further increase the temperature to $50°F$ ($10°C$). After allowing for a temperature fall in the evaporator, the limiting pressures and temperatures then become:

Evaporator $45.05°F$ ($3.8°C$) and 14 p.s.i.a. (allowing a $5°F$ drop)

Condenser $130°F$ ($54.5°C$) and 65 p.s.i.a.

$$\frac{p_1}{p_2} = \frac{65}{14} = 4.65$$

$q_2 = H_B - H_A = 125 - 22 = 103$ Btu/lb $\quad$ (see refrigeration tables or p/h chart)

$q_1 = H_C - H_D = 148 - 22 = 126$ Btu/lb $\quad$ (" "   " " " ")

$w = H_C - H_B = 148 - 125 = 23$ Btu/lb $\quad$ (" "   " " " ")

$$q_2 = \frac{103}{126} \times 38,000 = 31,061 \text{ Btu/h}$$

$q_1 = 38,000$ Btu/h

$q_R = \text{mass flow of refrigerant} = \dfrac{31,061}{103} = 301$ lbs/h = 5 lbs/min

$\text{I.H.P.} = \dfrac{5 \times 23}{42.42} = 2.71$

$\text{P.E.R.} = R_\eta = \dfrac{126}{23} = 5.48$

Piston displacement = $5 \times 3.65 = 18.25 \text{ ft}^3/\text{min}$

$$p_1/p_2 = \frac{65 \text{ p.s.i.a.}}{14 \text{ p.s.i.a.}} = 4.64$$

Substitution of Practical for Ideal Cycle Value

As explained earlier in the book, the practical cycle will be less efficient than the ideal cycle.

1. Pressure and Temperature Fall

These variations from the ideal have been allowed for in the heat exchangers. There will, however, be friction losses in pipes, etc., still to be allowed for.

2. Compressor Losses

The ideal, theoretical volumetric efficiency gives only an approximation to the true value. Leakage, clearance and polytropic instead of theoretical ideal adiabatic (i.e. isentropic) compression all lead to an increase in practical horsepower. The actual frictional losses of the compressor may provide a mechanical efficiency of, say, 90 per cent at full load but at, say, 50 per cent load, since the frictional losses remain constant, the mechanical efficiency would then fall to 81 per cent.

Data given by one American authority using F12 are as follows:

Compression Ratio	2	3	4	5	6
Compressor Mechanical Efficiency (%)	79	78	77	76	75

British compressor manufacturers offer a somewhat lower efficiency. In one case a 5 kW compressor provided the following figures:

Motor input 17,150 Btu/h (5815 W)

Adiabatic work 9500 Btu/h (2785 W)

Mechanical efficiency = 55.4 per cent

The effect of compressor losses, when using a British compressor (case (a)) would be to increase from an I.H.P. of 2.97 for the ground coil heat pump to 5.55 B.H.P. and to reduce the value of R_η from 5.03 to 2.97. If the efficiencies quoted above were substituted the B.H.P. = 4.3 and R_η is reduced from 5.03 to 2.72 (without sub-cooling).

The Effect of Recovering Heat Losses

The effect of capturing 90 per cent of the compressor and sensible heat losses can be worked out for case (a) as was shown for the oil engine driven heat pump and it would be to increase the temperature of the incoming antifreeze liquid by 7.7°F giving a higher evaporator pressure of 40 p.s.i.a. and a reduction of pressure ratio (r) from 5.25 to 4.8. This reduction in the value of 'r' has important effects as shown by the following data in Table 18, which assume the higher values of mechanical efficiency.

PROPERTY	CASE A (No sub-cooling)		CASE A (sub-cooling)		CASE B		OIL ENGINE	
	Btu/h	kW	Btu/h	kW	Btu/h	kW	Btu/h	kW
q_1	38,000	11.14	38,000	11.14	38,000	11.14	38,000	11.14
q_2	30,400	8.78	26,740	7.75	29,300	8.60	31,061	9.10
q_R	12.35 lbs/min		11.53 lbs/min		11.87 lbs/min		5 lbs/min	
I.H.P.	2.97		2.95		3.41		2.71	
Displacement	12.83 ft^3/min		12.5 ft^3/min		13.80 ft^3/min		18.25 ft^3/min	
p_1/p_2	5.25		5.1		6.13		4.64	
$R_{\eta(I)}$	5.03		5.45		4.37		5.48	

Practical Values

B.H.P. (ϵ = 76%)	4.07		3.9		4.6		3.56	
$R_{\eta(P)}$	3.65		3.83		3.23		4.17	

Table 18. Summary of findings.

In the practical machine there will be some further additional losses to be considered, e.g. the nature of the flow refrigerant and a variation from true isothermal working along lines AB and CD in Fig. 16 etc. There is still room for improvement in the design of a practical heat pump that will meet the value of P.E.R. = 3 specified in this book.

Appendix C   The "value" of heat

The value of most commodities is based upon the two factors of scarcity and quality. Energy in the form of heat appears to be an exception to this rule. Since we are approaching the end of the world's fossil fuel supplies, it is assumed that there is a condition of short supply to our sources of heat energy and this view is registered by the rapid increase that has taken place in fuel costs since the first quarter of this century. Yet heat energy never has and never will be in short supply; at least so far as useful low-grade heat is concerned. Because we have ignored the factor of quality, high-grade heat is expensive and low-grade heat is considered worthless. If we wish to provide 1 therm per hour of heated air at 100°F to maintain a building at 70°F it is possible to do this by destroying 12 lbs of oil fuel per hour to provide 1 therm/hour of expensive high-grade heat for that purpose. However, it is also possible to achieve the same result by burning only 6 lbs of fuel in an engine to provide 33,000 Btu/h of high-grade fuel heat and to supply the remaining 67,000 Btu/h as low-grade heat at 33°F obtained from earth or air. It is therefore necessary to distinguish between the quality, as well as the quantity, of heat used.

High-grade heat consists of heat which can suddenly be released at a high temperature and is usually locked up ready for release, as in coal, gas or oil. Further, once the heat energy is released from its locked form it must inevitably fall rapidly in quality until the temperature has fallen from its expensive high temperature to that of our ambient surroundings in earth, air or water. Like water, the heat falls on hills from which it inevitably descends to the plains and the sea.

The entire world's daily existence depends upon having these sources of high-grade, high temperature heat which can be released without effort on our part so as to fall to the surrounding ambient temperature and in so doing, heat our rooms and drive our engines. We, therefore, place a high value and cost on the locked-up high-grade heat and place no value on the almost illimitable quantity of de-graded ambient heat (which is no less in quantity than in its former state) which results from the heating and power operations, and which lies around us in an abundant sea.

But the time is fast approaching when the highly valued high-grade heat will no longer be available and then most of the world's daily activities may cease. We are being compelled to ask whether we can find a process which can utilise the huge quantity of low-grade heat which surrounds us, and in so doing heat our rooms by using a fraction of the rapidly diminishing sources of high-grade heat.

During the first half of this century the price of heat from coal or oil was grossly undervalued. As a result, the price of coal has increased by a factor of 100 since 1900 and that of oil by a factor of 15. The considerable price increases due to the rapid diminution of these sources of high temperature heat are leading to an examination of the potentialities of the low temperature heat existing in our environment. If we have soil at 47°F, surrounding and giving heat to, one lineal foot of thin pipe carrying liquid at 33°F, it will provide the same heat quantity (33.6 Btu or 9.8 W) as would be found by passing 10 lbs of warmed air at 82°F into a room which it is desired to maintain at 68°F.

Therefore, in differentiating between the quantity and quality of a given amount of heat we must consider, firstly, high-grade heat contained in fossil fuels which is in short diminishing supply and rapidly growing more expensive and, secondly, low-grade heat which is available in almost illimitable quantities in soil and air, and which, because it is at lower temperatures, is quite wrongly considered to be worthless.

To deliver 1 therm of heat at 100°F from fuel oil (at 50 per cent combustion efficiency) requires that 12 lbs (5.36 kgs) of fuel shall be irretrievably destroyed. If a heat pump ($R_{\eta} = 3$) is used to provide 1 therm at 100°F it is necessary to provide 1 therm of low-grade heat at 33°F and 3 x 0.333 therm of power/heat equivalent from oil ($\eta = 0.33$) thus consuming 6 lbs of oil. The cost of 2 therms of oil heat for direct combustion is approximately £0.58 (2p per kW) whereas the alternative cost is £0.29 (1p per kW). The value of a commodity is defined as "its ability to exchange for, or preserve the use of, another commodity". It would seem that the provision of 1 therm of low-grade heat has effected a monetary saving of £0.29 and that this low-grade heat has a value of £0.29 per therm or 1p per kW.

We spend many millions of pounds annually on testing for new sources of coal or oil which will have a value of 1p per kW and, literally, nothing on the exploiting of the inexhaustible sources of ambient heat which have also got a value of 1p per kW. A recent study which was carried out by a research team at The University of East Anglia under the direction of Professor Vine and Dr Tovey on a buried coil in my garden has been carried out at our own expense and without the benefit of any grant or financial assistance. Apart from the research on ground coils carried out thirty years ago by Miss Griffiths, under the auspices of the British Electrical and Industrial Research Association, I have no knowledge of any other research on this important subject having been carried out in this country.

It should be noted that when ambient heat is taken from soil and then raised in temperature by doing work on it in the heat pump, we are only temporarily abstracting the heat. The high temperature heat (of a quantity greater than originally abstracted) will escape via walls and roof as it falls in temperature and thus inevitably will return to the source from which it came - but with an addition from the fuel used to drive the heat pump.

Appendix D   Postscript

The challenge of the OPEC countries in 1973 has had the same effect upon the world as that portrayed by Dr. Johnson on the man due to be executed in a fortnight in that "it concentrated the mind wonderfully".

Between February and June 1976 there have been more national conferences held to discuss heat pumps than have taken place in the previous thirty years. At one of these three conferences, in Oxford, the views of the Electricity Supply Industry regarding electrically driven heat pumps of 3 kW demand, capable of heating a complete small house, were expressed as follows:

Problem: "The E.S.I. will not actively encourage the use of heat pumps on a 24 hour availability."

Solution: "Investigate thermal storage facilities to provide beneficial characteristics to the E.S.I. . ."

The distinction which the E.S.I. makes between heat pumps with a total maximum demand of 3 kW and electric fires and cookers with demands of 3 kW and 12 kW respectively is difficult to understand. Such electric fires or cookers are very "actively encouraged" by them without any restriction for use "on a 24 hour availability". This is also true of electric block heater loads and 'electricaire' house heating systems of 12 kW or more, which could be replaced more efficiently by a 4 kW heat pump running for the same periods.

Apart from action taken by the two Midland Electricity Boards, there is very little reason to assume that the E.S.I., in its present form, will adopt the solution recommended in June 1976 at Oxford, having ignored the same solution placed before it a quarter of a century ago. A somewhat more advanced attitude is shown by the R.W.E., the largest electricity supply authority in Germany, as is shown by the following extract from "Refrigeration and Air Conditioning" published in June 1976.

"An experimental house built at Aachen in West Germany by the Philips company is heated by a solar collector and earth source heat pump. By using a high standard of insulation and reclaiming heat from the exhaust air and waste water, energy requirements of the house have been cut to one-fifth of the normal.

"The project is part of a £2m., three year investigation into improved heat conservation, financed partly by the Government and partly by Philips, and RWE – the largest electrical supply authority in Germany. Although the house itself has cost around £100,000 to build, a very large share has gone into the metering equipment, the solar collector, and the ground coils as well as the services installations such as the heat pump, hot water recovery units and water storage vessel. At the moment the house is being run with simulated occupancy, the various heat sources and activities being generated by computer; later on, around mid-1977, the house will be occupied by a family of four.

"Similar in appearance on the outside to many houses in small German towns, the Philips house has only the ground floor as the living zone. Part of the steeply sloping roof space is taken up by the solar collector, and part is used for the data-logging equipment, and the Philips 855 process control computers.

"Fresh air is supplied through a balanced mechanical ventilation system giving about one air change per hour in the living space in winter, and two air changes in summer, corresponding to flow rates of 180 and 360 CFM. Air heating load would have been as much as 24,000,000 Btu's per year, but a rotary regenerative heat exchanger reduces this to only 4.8m. Btu's.

"Primary source for the heat pump is a 400 ft long ground coil placed under the basement floor and covering an area of about 1500 sq.ft. Source temperature in winter is reckoned to be a constant 45°F and this is pumped up to about 120°F and transferred to the heat exchanger. Performance energy ratio is expected to be about 3.5, and calculations have shown that the ground coil will be capable of heating the whole house in winter.

"The heat pump will also be used to recover heat from waste water by means of a coil in the 220 gallon waste water accumulator tank. From the normal domestic water, plus the output of the washing machine and dish washer, the heat pump will save about 3000 kWh per year out of the potential loss of 3160 kWh.

"Another phase of the experiment is to turn to the solar collector as the only source of heat. By using

How the thermal performance of the Philips house compares with traditional designs.

	Average house		Well insulated house		Philips house	
	U value	MBtu's year	U value	MBtu's year	U value	MBtu's year
Walls, floor ceiling	0.22	111	0.08	43	0.02	12
Glazing	1.0	34	0.58	19	0.26	31
Ventilation		24		24		7*
Total		169		86		28

* Includes 2 MBtu's for uncontrolled infiltration

Structural details:

Floor area 1250 sq ft
Glazing area 250 sq ft
Living space 10,000 cu ft

(Table based on Philips experimental data)

a concentrating type of collector made up of banks of partly silvered, evacuated glass tubes containing the pipe grids, Philips estimates an annual efficiency of about 50 per cent., yielding 10,000 to 12,000 kWh in an average year from an area of 200 sq ft. The ability of such collectors to produce water at temperatures approaching 220°F is important in cooling applications where absorption machines have to be used.

"Essential to the success of both solar and ground sources is the 9000 gallon water storage reservoir housed in the basement.

"Summer cooling requirements of the Philips house will be small, mainly because of the low solar gain through structure and glazing. However, the designers intend taking advantage of natural cold storage in the earth by leading the fresh air through a clinker block shaft next to the basement. They expect conduction cooling from the surrounding earth to reduce inlet air temperatures to 64°F, giving an equivalent cooling rate of 10,000 Btu's/hr.

"The yearly total for the structural losses in the Philips house is only 21 MBtu's compared with a possible 145 MBtu's had traditional insulation standards been used. The heat recovery wheel on the mechanical ventilation system reduces the fresh air load from a possible 24 MBtu's to only 7 MBtu's per year. Thus, compared with the traditional house, the Philips design reduces the annual total load from 169 MBtu's to 28 MBtu's (see table).

Bibliography

1. Europe's Growing Needs of Energy. O.E.E.C. 1956

2. Energy Supplies During the Next Fifty Years. J. Sumner. I.E.E. Journal. Vol.3 1957.

3. The Norwich Heat Pump. J.A. Sumner. Journal I.Mech.E. Vol. 158 No.1, June 1948.

4. Domestic Heating by Heat Pumps. J.A. Sumner. Journal I.H.V.E., June 1955.

5. Economy of Heating or Cooling Buildings. W. Thomson. Glasgow Plub. Soc. Proc., Vol. 111 Dec.1852.

6. Heating and Refrigeration of Air. W. Thomson. Cambridge Mathematical Journal 1852.

7. Heat Pumps and Thermo-Compressors. S.J. Davies.D.Sc. Constable and Co. 1950.

8. Heat Pump Applications. Kember and Oglesby. McGraw-Hill 1950.

9. The Heat Pump – Practical Application. J.B.Pinkerton. Princes Press, London. 1949.

10. Refrigerating and Heating Journal, July 1976.